mL

UNIVERSITY OF STRATHCLYDE

30125 00503785 7

KV-038-653

Books are to be returned on or before
the last date below.

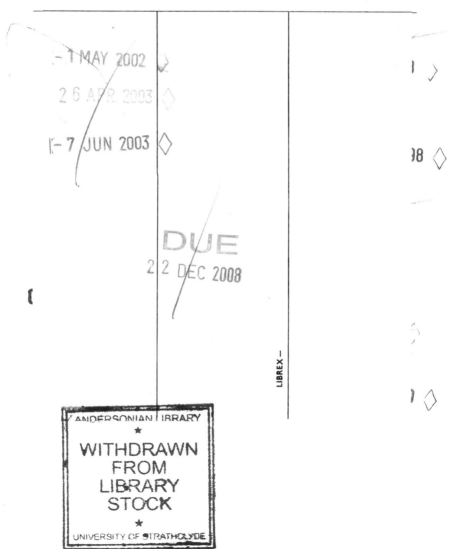

- 1 MAY 2002

2 6 APR 2003

- 7 JUN 2003

DUE

2 2 DEC 2008

LIBREX —

ANDERSONIAN LIBRARY
★
WITHDRAWN
FROM
LIBRARY
STOCK
★
UNIVERSITY OF STRATHCLYDE

Biotechnology for the Treatment of Hazardous Waste

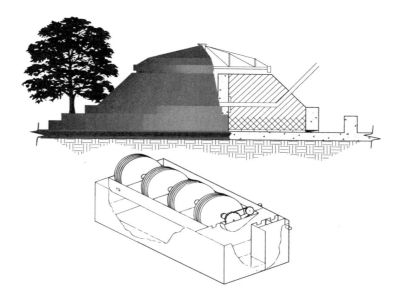

Edited by

Daphne L. Stoner

LEWIS PUBLISHERS
Boca Raton   Ann Arbor   London   Tokyo

UNIVERSITY OF STRATHCLYDE

1 7 AUG 1995

UNIVERSITY LIBRARY

Library of Congress Cataloging-in-Publication Data

Biotechnology for the treatment of hazardous waste / edited by Daphne L. Stoner.
    p. cm.
    Includes bibliographical references (p.) and index.
    ISBN 0-87371-613-2
    1. Hazardous wastes—Biodegradation. 2. Biotechnology.
    I. Stoner, Daphne L.
TD1061.B54   1993
628.4′2—dc20
                                                      93-23166
                                                     CIP

This book contains information obtained from authentic and highly regarded sources. Reprinted material is quoted with permission, and sources are indicated. A wide variety of references are listed. Reasonable efforts have been made to publish reliable data and information, but the author and the publisher cannot assume responsibility for the validity of all materials or for the consequences of their use.

Neither this book nor any part may be reproduced or transmitted in any form or by any means, electronic or mechanical, including photocopying, microfilming, and recording, or by any information storage or retrieval system, without prior permission in writing from the publisher.

All rights reserved. Authorization to photocopy items for internal or personal use, or the personal or internal use of specific clients, may be granted by CRC Press, Inc., provided that $.50 per page photocopied is paid directly to Copyright Clearance Center, 27 Congress Street, Salem, MA 01970 USA. The fee code for users of the Transactional Reporting Service is ISBN 0-87371-613-2/94/ $0.00+$.50. The fee is subject to change without notice. For organizations that have been granted a photocopy license by the CCC, a separate system of payment has been arranged.

CRC Press, Inc.'s consent does not extend to copying for general distribution, for promotion, for creating new works, or for resale. Specific permission must be obtained in writing from CRC Press for such copying.

Direct all inquiries to CRC Press, Inc., 2000 Corporate Blvd., N.W., Boca Raton, Florida 33431.

© 1994 by CRC Press, Inc.
Lewis Publishers is an imprint of CRC Press

No claim to original U.S. Government works
International Standard Book Number 0-87371-613-2
Library of Congress Card Number 93-23166
Printed in the United States of America 1 2 3 4 5 6 7 8 9 0
Printed on acid-free paper

D
628.54
BIO

Dedication

The editor wishes to dedicate her efforts in writing and compiling this book to Dr. W. Murray Bain.

Editor

Daphne L. Stoner, Ph.D., is a microbiologist with the Biotechnology Group at the Idaho National Engineering Laboratory, Idaho Falls, Idaho. In 1975 Dr. Stoner received a B.S. degree from the University of Maine, Orono and in 1986 a Ph.D. from the University of Maryland, College Park. She joined the staff at the Idaho National Engineering Laboratory originally as an Associated Western Universities postdoctoral fellow and later as a staff scientist. Research interests include: the development of intelligent control systems for mixed-culture bioprocessing; degradation of organosulfur compounds; modification of organic sulfur functionalities in coal and coal-derived materials; and the development of on-line or real time molecular methods for the analysis of mixed microbial communities.

Preface

The development of biologically-based processes for the treatment of hazardous wastes is a multi-disciplinary effort requiring the consideration of a number of biological, chemical, and physical parameters as well as the effective teaming of biologists, chemists, engineers, and regulatory agencies. To successfully work at the interface of merging disciplines requires an appreciation of the fundamentals of the participating professions. While this book is far from a comprehensive examination of all the factors that need to be considered for the development of biologically-based treatment processes, it bridges the disciplines in such a manner that an exchange of fundamental information can take place.

While hazardous waste can contain inorganic and organic constituents, application of biological processes to the treatment of organic hazardous waste has been the subject of far more symposia, books, and articles than the treatment of inorganic waste. This book attempts a balance between inorganic and organic hazardous wastes. In Chapter 1, the organic hazardous waste amenable to biological treatment is explained by summarizing the metabolic transformations that enable microorganisms to degrade organic compounds. A similar approach was taken by Henry L. Ehrlich in Chapter 2 to describe the inorganic hazardous waste amenable to biological treatment. Dr. Ehrlich continues with examples of inorganic hazardous waste treatment with suggestions for future applications.

The information provided in Chapters 1 and 2 is an overview of the fundamental processes upon which biologically-based treatment processes may be developed. However, the physical and chemical characteristics of the waste as well as the physiology of the microorganisms control the methods and approaches used for waste treatment. Daphne Stoner, in Chapter 3, describes the engineering approaches and strategies utilized for the treatment of organic waste forms. In Chapter 4, Douglas Gould and Ron McCready discuss the treatment strategies that are in use or proposed for the treatment of a variety of inorganic wastes. Many processes developed for the recovery of metals for the biohydrometallurgical industry have application to the treatment of inorganic hazardous waste.

For waste treatment processes which require live, respiring biomass, certain parameters must be maintained in order to sustain microbial activity. Furthermore, the type of waste and the characteristics of the surrounding environment often impact treatment options or approaches. In Chapter 5, Peter Adriaens and William Hickey describe the physiological parameters that must be considered when applying microbial processes to waste treatment or environmental remediation. Advances in the area of biotechnology have application to improve treatment processes. Burt Ensley, in Chapter 6, discusses the fundamental approaches to improving the characteristics of microorganisms in order to enhance their waste treatment capabilities.

In Chapter 7, Graham Andrews discusses the fundamentals of process engineering as it applies to the design of the biologically-based hazardous waste treatment systems. Limitations to process rates are discussed. It is in this chapter that some of the fundamental differences between the engineering and scientific disciplines are apparent. Sue Markland Day describes, in Chapter 8, the influence of environmental regulations and mandates on the development and implementation of waste treatment. Environmental regulations, while imposing limits on treatment technologies and options, has stimulated the market for biotechnological applications to waste treatment and environmental remediation.

Biologically-based waste treatment and environmental remediation technologies are currently being developed and applied in a worldwide effort to offer suitable alternatives to the conventional treatment technologies. While new information is continually being gathered, the foundation for biologically-based waste treatment and remediation technologies has already been established. Biodegradation of hazardous organic compounds has been demonstrated; degradative pathways determined, and the physiological requirements of the microorganisms understood. Similarly, the accumulation and transformation of hazardous inorganic materials are widely recognized, and the active and passive processes involved in the inorganic transformations understood. These fundamentals have been applied to the development and implementation of full-scale bioremediation and waste treatment systems. What remains is the continued and improved application of the fundamental information and the experience gained from field operations to the development of processes that offer permanent and effective solutions to waste treatment and environmental remediation problems.

Contributors

Peter Adriaens, Ph.D.
Visiting Assistant Professor
Department of Civil and
  Environmental Engineering
University of Michigan
Ann Arbor, Michigan

Graham Andrews, Ph.D.
Principal Engineer
Center for Biological Processing
  Technology
Idaho National Engineering
  Laboratory
Idaho Falls, Idaho

Sue Markland Day, B.A.
Senior Research Associate
Waste Management Research and
  Center for Environmental
  Biotechnology
University of Tennessee
Knoxville, Tennessee

Henry L. Ehrlich, Ph.D.
Professor
Department of Biology
Rensselaer Polytechnic Institute
Troy, New York

B. D. Ensley, Ph.D.
Director
Envirogen
Princeton Research Center
Lawrenceville, New Jersey

W. D. Gould, Ph.D.
Biohydrometallurgist
CANMET
Department of Energy, Mines, and
  Resources
Ottawa, Ontario, Canada

William J. Hickey, Ph.D.
Department of Soil Science
University of Wisconsin
Madison, Wisconsin

R. G. L. McCready, Ph.D.
Director
Scientific and Technical Services
Industrial Biotechnology Services Inc.
Calgary, Alberta, Canada

Daphne L. Stoner, Ph.D.
Senior Scientist
Idaho National Engineering
  Laboratory
Environmental Biotechnology,
  EG & G Idaho, Inc.
Idaho Falls, Idaho

Contents

CHAPTER 1

Hazardous Organic Waste Amenable to Biological Treatment

Daphne L. Stoner

1. INTRODUCTION

The diverse metabolic and physiological characteristics of microorganisms, in addition to their ability to thrive in a variety of environments, may be exploited for the purposes of environmental remediation and waste treatment. Axenic cultures or mixed populations of microorganisms with the ability to degrade or mineralize hazardous materials can form the basis of a bioprocess for the treatment of organic hazardous waste or serve as a refinement of many conventional waste treatment processes. Research has progressed from "bench scale studies" to field and treatability studies for the remediation of contaminated sites.[1,2] In many instances, bioremediation is being actively pursued as the preferred treatment option.[3]

0-87371-613-2/94/$0.00+$.50
© 1994 by Lewis Publishers

2. PHYSIOLOGICAL BASIS FOR HAZARDOUS ORGANIC COMPOUND DEGRADATION

A survey of the literature reveals a wealth of information concerning the degradation of a diverse array of organic compounds by a variety of microorganisms as well as reports of the degradation of organic chemicals in the environment by a combination of biotic and abiotic processes. Advancement in biologically based treatment strategies for hazardous waste treatment and environmental remediation is, in part, dependent upon understanding the mechanisms by which organisms degrade organic materials. The intent of this chapter is not an exhaustive review of this literature, but rather a discussion of the features common to the microbial degradation of organic compounds and their implication for hazardous waste treatment and environmental remediation.

Microbial transformations applicable to waste treatment are those reactions that microorganisms mediate to satisfy nutritional requirements, satisfy energy requirements, detoxify their immediate environment, or are the indirect or unexpected result of metabolic processes or the physical or chemical characteristics of the microbial cell. While many transformations applicable to waste treatment are of a direct benefit to the microorganisms, some transformations are the result of fortuitous reactions that do not offer any advantage to the microorganisms.

In order to satisfy their nutritional and energy requirements, heterotrophic microorganisms metabolize a variety of organic compounds. Anabolic metabolism results in the synthesis of a diverse array of monomers, polymers, and complex macromolecules from simple precursor molecules. Often the precursor molecules are derived from the degradation or catabolism of other, frequently more complex, organic molecules. Important to the microbial degradation of hazardous organic materials are the catabolic or degradative capabilities that enable microorganisms to degrade these compounds and utilize them as suitable substrates for growth.

The transformations resulting in the degradation of organic materials can be classified into two broad categories: mineralization and cometabolism. Whether a compound is mineralized or cometabolized has implications for the development of a waste treatment process or for environmental remediation. Mineralization is the complete conversion of organic materials to inorganic products. Mineralization is a growth-linked process that involves the central catabolic and anabolic pathways of a microorganism.[4] A compound that is mineralized serves as the growth substrate and energy source for the microorganism. In general, only a portion of the organic compound is incorporated into cell material with the remainder forming metabolic by-products such as CO_2 and H_2O. Mineralization can also occur by the combined activities of a microbial consortia.[5,6] Interdependent members of the community are required to effect the complete conversion of the hazardous material to satisfy growth requirements.

Cometabolism is the degradation of organic compounds usually via nonspecific enzymatically mediated transformations.[4] In contrast to mineralization, cometabolism does not result in the increase in cell biomass or energy. Consequently, the ability to cometabolize a compound is not a benefit to the microorganism. In fact, another substrate is necessary in order to satisfy growth and energy requirements of the cell. Typically, cometabolism results in the modification or transformation of the organic material and does not result in the complete destruction of the molecule. While cometabolism can result in the complete destruction of an organic molecule, the accumulation of potentially toxic intermediates in the environment can occur.

3. ANAEROBIC VERSUS AEROBIC TRANSFORMATIONS

Suitable electron donors and electron acceptors are required for the energy yielding reactions of the cell. Microorganisms that utilize organic substrates as the source of carbon for growth typically utilize the organic substrate as the electron donor (energy source) as well. Depending on the microorganism, a variety of inorganic species and organic molecules can serve as electron acceptors. Aerobic microorganisms utilize oxygen as the terminal electron acceptor and are limited to environments that contain sufficient amounts of oxygen. Anaerobic microorganisms are those microorganisms that do not utilize oxygen as the terminal electron acceptor. Anaerobic microorganisms use a range of electron acceptors, which, depending on the redox conditions and availability, include nitrate, iron, manganese, sulfate, and CO_2.

Hazardous organic materials can be biodegraded under aerobic or anaerobic conditions. The type and form of the organic hazardous material, the catabolic capabilities of the microorganisms, and the availability of nutrients and electron acceptors may dictate whether aerobic or anaerobic degradation is the preferred treatment option.[7] In cases where there is little or no oxygen or there is an abundance of an alternative electron acceptor such as nitrate or sulfate, degradation of hazardous organic materials by anaerobic microorganisms should be considered. The degradation of organics by anaerobic bacteria is of importance when remediating anoxic environments such as saturated subsurface soils, landfills, lagoons, and some groundwaters. The subject of anaerobic degradation of organics is discussed in more detail in Chapter 5 of this volume.

4. EXAMPLES OF HAZARDOUS WASTE AMENABLE TO BIOLOGICAL TREATMENT

Organic compounds that can be metabolized by microorganisms are amenable to treatment by waste treatment bioprocesses. The mineralization of organic chemicals by microorganisms or microbial consortia that utilize these compounds for their carbon and energy source assures the complete destruction of these hazardous materials. Hence, mineralization of organic materials is preferred for the microbial treatment of organic waste or environmental remediation as it potentially offers a permanent solution. Microbially based treatment processes are suitable for concentrated wastes as well as dilute aqueous based wastes. Thus, microbial degradation is perceived as a low cost alternative to incineration of concentrated waste forms as well as other oxidative treatments such as the photooxidation of dilute waste streams.

Partial modification of hazardous organics is not considered an environmentally sound treatment option unless another microbial population is present to complete the destruction. For example, partial degradation of trichloroethylene can lead to the accumulation of chlorinated intermediates such as dichloroethylene or vinyl chloride. In essence, the problem of trichloroethylene contamination can be exchanged for dichloroethylene or vinyl chloride contamination unless another population is present that can degrade these toxic intermediates. For this reason, transformations that lead to the mineralization of organic materials by axenic cultures or microbial consortia will be emphasized in the remainder of this chapter.

Whether a compound is mineralized anaerobically or aerobically, a review of the degradation pathways reveals the fact that many compounds are degraded via common metabolic intermediates. It is the peripheral transformations that transform diverse compounds to common intermediates. Within a class of compounds, such as aromatic hydrocarbons, features common to the catabolism of the seemingly different compounds are readily apparent.

The description of pathways applicable to hazardous waste treatment will begin by a discussion of the degradation of aliphatic and aromatic compounds found in petroleum. The section will continue with a discussion of the transformation of halogenated compounds. Halogenated compounds, once dehalogenated, are degraded by the pathways involved with aromatic or aliphatic hydrocarbon degradation.

4.1 Petroleum Components

The microbial degradation of petroleum, the taxonomy and distribution of petroleum-degrading microorganisms, the chemical and physical factors that affect biodegradation, and the application of biotechnology to the remediation of petroleum contaminated environments are the subjects of numerous articles and books.[8-14] Information on biodegradation may often be found in books concerning the remediation of oil-contaminated environments.[15,16]

Petroleum is degraded by taxonomically diverse bacteria and fungi from terrestrial,[8,9,11,17] freshwater,[8,9,18] and marine environments.[8,9,19] The metabolic pathways involved with the degradation of petroleum components have been well characterized. Central catabolic pathways in microorganisms modify the seemingly diverse alkane and 2-3 ring aromatic compounds present in petroleum to their common metabolic intermediates. Depending on their chemical nature, petroleum components are degraded by mineralization or cometabolism.

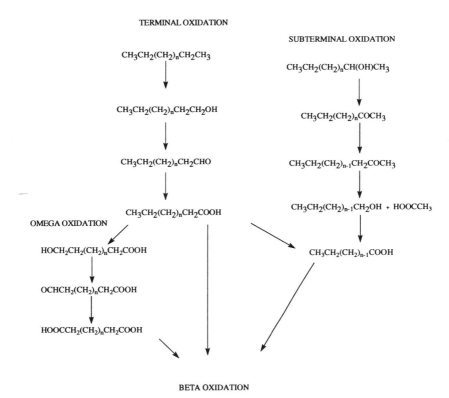

Figure 1. Degradation of aliphatic hydrocarbons by aerobic microorganisms by omega-oxidation, terminal oxidation, and subterminal oxidation. Redrawn from Watkinson and Morgan.[12]

4.1.1 Alkanes

Normal alkanes can be degraded by monoterminal, diterminal (ω-oxidation), or subterminal oxidation (Figure 1). The biodegradation of *n*-alkanes, however, is generally accepted to be initiated by the monoterminal modification of the alkane into a primary alcohol, followed by sequential oxidation to the corresponding aldehyde and monocarboxylic acid.[8,12,20] Degradation of the carboxylic acids proceeds via β-oxidation with the formation of acetate units as acetyl coenzyme A. Thus, the carboxylic acids are shortened, successively, by two carbons. The acetate units cleaved from the alkanes are metabolized via the central metabolic pathways with the eventual release of CO_2. While branching does increase the resistance of alkanes to microbial attack, degradation can proceed via alternative pathways such as α-oxidation, ω-oxidation, or β-alkyl group removal. The degradation of branched alkanes via ω-oxidation results in the formation of dicar-

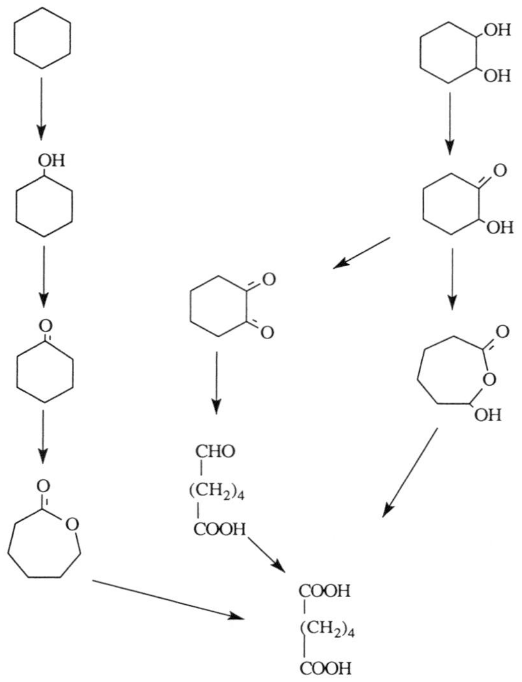

Figure 2.   Degradation of cyclic aliphatic hydrocarbons. Compiled from Perry.[21]

boxylic acids. Fatty acid profiles of bacteria influenced by alkane used for growth are evidence that degradation of the alkane need not occur exclusively by β-oxidation and complete degradation by central metabolic routes.

Unsubstituted cyclic alkanes, major components of crude oil, are relatively resistant to microbial degradation and tend to persist in the environment. Even though cyclic alkanes are widespread in the environment due to natural processes and human activities, relatively few microorganisms are able to utilize cycloalkanes for growth.[21-24] However, the cometabolism cyclic alkanes are apparently widespread. The resulting oxygenated derivatives, cycloalkanols and cycloalkanones, can serve as growth substrates for other microorganisms. Thus, the degradation of cyclic alkanes by mixed populations in the environment is thought to be initiated by cometabolism and followed by mineralization of cometabolites.

Cyclic alkanes are oxidized to the corresponding alcohols or ketones (Figure 2). Degradation then proceeds via ring cleavage. For example, cyclohexane is metabolized via cyclohexanol, cyclohexanone, and ε-caprolactone to adipic acid by *Xanthobacter* sp.[24] Cyclohexanol degradation occurs by a similar metabolic pathway. In contrast to cyclic alkanes, alkyl substituted cyclic alkanes serve as growth substrates for microorganisms. The mode of degradation may be via β-oxidation of the alkyl side chain or via the oxidation of the ring.

Anaerobic degradation of hydrocarbons does occur. A sulfate-reducing bacterium, tentatively identified as *Desulfobacterium oleovorans*, was able to utilize

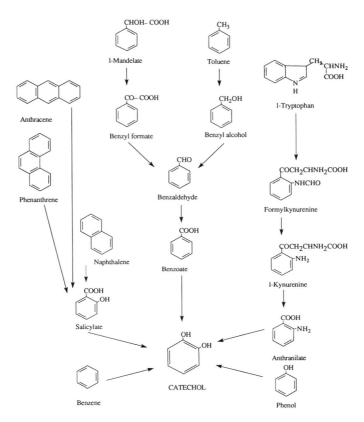

Figure 3. Generalized scheme for the aerobic degradation of aromatic hydrocarbons to catechol. Redrawn from Stanier and Ornston.[28]

C12 to C20 alkanes for growth.[25] Based on the amount of hexadecane degraded and sulfide produced, degradation was presumed to have occurred through an oxidative mechanism with hexadecane being completely converted to CO_2.

4.1.2 Aromatic Hydrocarbons

The aromatic compounds found in fossil fuels can occur with alkyl substituents or with heteroatomic nuclei containing oxygen, nitrogen, or sulfur. Microorganisms that utilize aromatic compounds for growth are widely distributed in nature and have the metabolic capability of degrading condensed aromatics and alkylated derivatives, which range from one ring compounds such as benzene and toluene to 4 or 5 ring compounds such as benz(a)anthracene and benzo(a)pyrene, respectively.[13,26,27] Figure 3 presents a generalized scheme for the microbial modification of aromatic and polycyclic aromatic compounds degraded through common metabolic intermediates.[28] Initial oxidation by bacteria involves the incorporation of two atoms of molecular oxygen by a dioxygenase enzyme and the formation of a *cis*-dihydrodiol. Further oxidation leads to the formation of catechols that

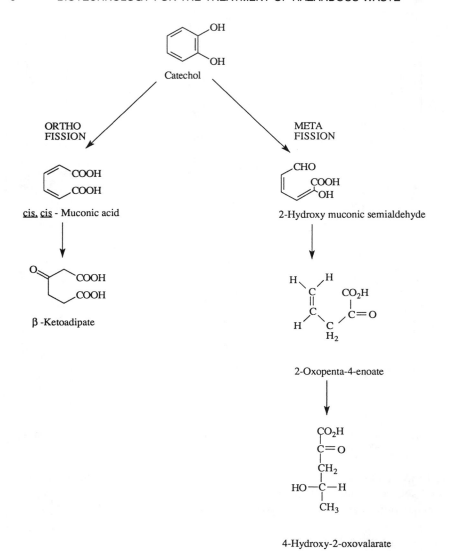

Figure 4. Aromatic ring cleavage pathways of aerobic bacteria. After Smith;[13] Stanier and Ornston;[28] and Nelson et al.[63]

enables ring fission by *meta* cleavage or *ortho* cleavage pathways to occur. A generalized scheme for *ortho* or *meta* cleavage of aromatic ring is shown in Figure 4. *Ortho* cleavage involves the cleavage of the aromatic ring between the two adjacent hydroxyl groups, while *meta* cleavage occurs between a hydroxyl group and the adjacent carbon. Further modification of the common intermediates results in the formation of central pathway metabolites.[26] Eucaryotic microorganisms such as fungi and yeast mediate the initial oxidation of aromatic compounds

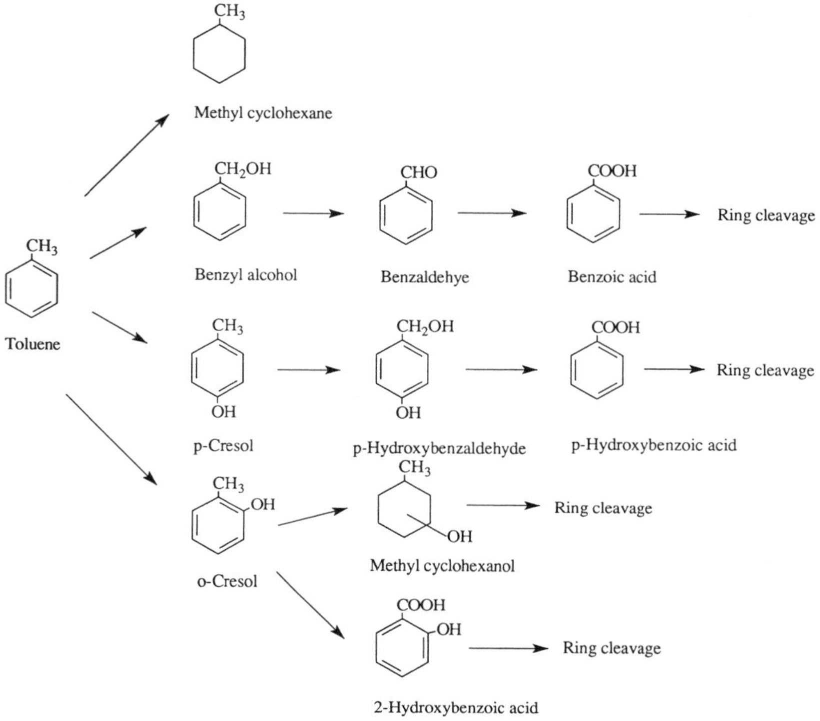

Figure 5. Anaerobic degradation of aromatic hydrocarbons. Redrawn from Grbić-Galić.[31]

with a cytochrome P450 catalyzed monooxygenase and epoxide hydrolase results in the formation of *trans*-dihydrodiols.[27]

Degradation of aromatic hydrocarbons by anaerobic microorganisms has been examined, and extensive reviews of this information are available.[29-32] Ample evidence exists for the microbial degradation of aromatic compounds under fermentative,[29-31,33] denitrifying,[29-31,34-36] sulfate-reducing,[30,37-39] ferric iron-reducing,[40,41] and methanogenic conditions.[29-31,42] While methanogenic consortia are known to degrade aromatic compounds, the presence of fermentative microorganisms is essential to degrade the organic compounds to the simple substrates used by the methanogens.

Benzoic acid intermediates are formed during the anaerobic degradation of toluene by nitrate-reducing, sulfate-reducing, and methanogenic cultures.[5,31,34,35,43,44] Initial attack on toluene (Figure 5) is most frequently postulated to occur by: (1) the hydroxylation of aromatic ring to form *para* cresol, which then can be converted to *p*-hydroxybenzoate and benzoate,[31] or (2) the hydroxylation of the methyl group to form benzyl alcohol.[30,31,45] Similarly, an initial oxidation of benzene involving the incorporation of oxygen from water results in the formation of phenol during anaerobic fermentation.[33]

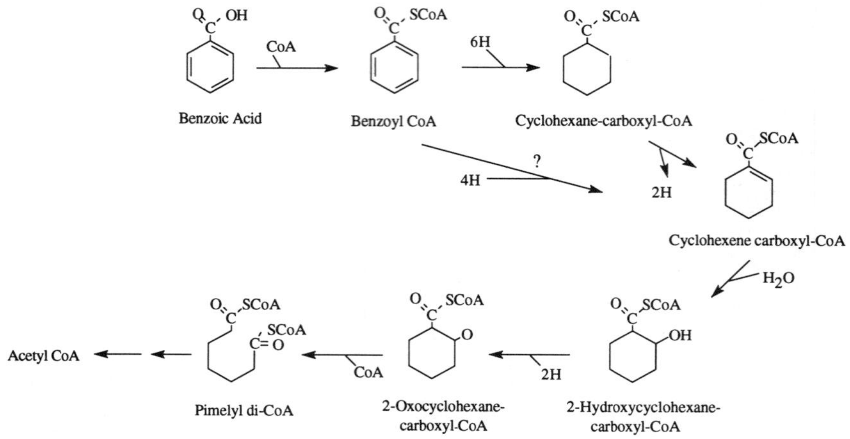

Figure 6. Degradation of benzoate by anaerobic photocatabolism. Redrawn from Geissler et al.[83]

A facultative bacterium designated strain T1[35] and a sulfate-reducing enrichment culture[43] appear to degrade toluene by two metabolic pathways. The pathway that leads to the mineralization of toluene involves acetyl coenzyme A. The other pathway that leads to the formation of dead-end metabolites involves succinyl-coenzyme A.[34–36]

Of interest is the anaerobic oxidation of aromatic compounds by an iron-reducing bacterium.[40,41] The bacterium, designated GS-15, was able to utilize toluene, phenol, or p-cresol as the sole carbon source with Fe(III) as the terminal electron acceptor. Because of the transient accumulation of p-hydroxybenzoate during growth on phenol or p-cresol, the initial step in the metabolism of phenol was suggested to be a carboxylation step. A carboxylation step was also postulated for the anaerobic degradation of phenol by denitrifying pseudomonads.[46] The metabolism of p-hydroxybenzylalcohol and p-hydroxybenzaldehyde by GS-15 suggests that these compounds are intermediates in the metabolism of p-cresol.

It is assumed that the degradation benzoate intermediates formed during the catabolism of aromatics proceed by the pathways described in earlier studies. Benzoate metabolism by anaerobic photocatabolism[47,48] is initiated by the formation of the coenzyme ester (Figure 6). The aromatic ring is further modified via a reductive pathway and hydroxylation by ring hydration. However, for a nitrate-reducing culture, *Moraxella* sp., benzoate was metabolized without the formation of coenzyme A esters.[49] Instead, a reductive pathway for benzoate degradation was proposed (Figure 7).

4.2 Halogenated Compounds

Halogenated compounds can be utilized as a growth substrate or cometabolized by anaerobic and aerobic microorganisms and consortia. Degradation of haloge-

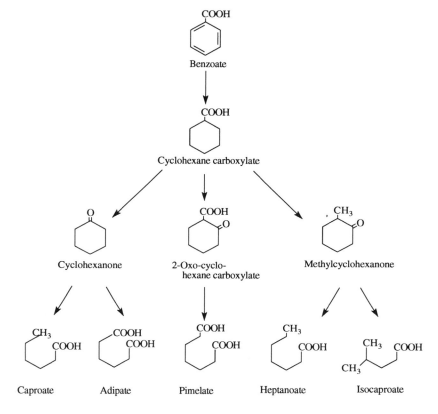

Figure 7. Anaerobic degradation of benzoate by methanogenic consortia. Redrawn from Grbić-Galić and Young.[42]

nated compounds can occur through the combined activities of fortuitously acting enzymes present in one or more microorganisms. The degradation of halogenated compounds has been summarized in several review articles that will assist the reader in summarizing the seemingly vast amount of information.[50-52] As with degradation of petroleum components described above, there are features common to the degradation of halogenated compounds. For instance, within a class of halogenated compounds, i.e., aliphatic, aromatic, and polycyclic aromatic, recalcitrance to microbial degradation is related to the number, type, and position of the halogen species on the molecule.[50,51] The degree of recalcitrance is related to the electronegativity of the substituent. That is, the recalcitrance of the carbon-halogen bond decreases in the order of F>Cl>Br>I. Dehalogenation can be spontaneous as in the loss of halogens during ring cleavage or enzymatically catalyzed reactions such as hydrolytic cleavage or reductive dehalogenation. In some cases, dehalogenation is the result of fortuitous reactions. Dehalogenation can also be the indirect result of microbial activity such as reductive dehalogenation of trichloroethylene to vinyl chloride by the sulfide generated by sulfate reducing bacteria.

If utilized for growth, metabolism through central catabolic pathways that have been described above for petroleum components is assumed once dehalogenation has occurred.

4.2.1 Halogenated Alkanes and Alkanoic Acids

Haloalkanoic acids, widely dispersed into the environment through their use as herbicides, also appear as intermediates arising from the degradation of more complex halogenated compounds. Haloalkanoic acids are readily degraded and utilized for growth by a variety of microorganisms.[53-55] As with other halogenated compounds, the degradation of the haloalkanoic acids is dependent on the halogen substituent and the regio- and stereospecificity of the enzyme. The halogen substituents are removed by hydrolytic cleavage, which results in the formation of hydroxyalkanoic acids from monosubstituted alkanes and oxoalkanoic acids from disubstituted compounds.[50] Once the compound is dehalogenated, the alkanoic acid is degraded and assimilated via central metabolic pathways.

The degradation of haloalkanes of short (C1-C4) and intermediate (C5-C9) chain lengths is of particular interest due to health and environmental issues. Reductive dehalogenation is the mechanism by which anaerobic microorganisms initiate the transformation of halogenated alkanes.[56] Under methanogenic conditions, reductive transformation of perchloroethylene (PCE) via trichloroethylene (TCE) and dichloroethylene (DCE) to vinyl chloride (VC) has been proposed by Vogel and McCarty,[57] and further transformation to ethylene by reductive transformation has been reported by Freedman and Gossett.[58] Figure 8 summarizes the degradation of chlorinated solvents by reductive dehalogenation. Transformation of PCE to ethylene in the absence of methanogenesis by acetogenic cultures suggested the involvement of a carbon monoxide-acetyl coenzyme A pathway in the dehalogenation.[59] Transformation to subsequently less halogenated derivatives proceeds at relatively slower rates. While the reductive dechlorination of chlorinated ethylenes by methanogens often leads to the production of VC, the partially dechlorinated products of PCE are more amenable to aerobic transformation.

Dehalogenation by oxygenase enzymes, glutathione-dependent dehalogenases and haloalkane halidohydrolases, is the mechanism by which the degradation of halogenated alkanes is initiated by aerobic microorganisms.[50] Oxidative dehalogenation is mediated by cytochrome P450 or other monooxygenases, while nucleophilic substitution is mediated by glutathione-S-transferases and results in the formation of aldehydes. Hydrolytic dehalogenation by haloalkane halidohydrolases yields alcohols as products. Aerobic transformation of halogenated alkanes is not as prevalent as aerobic dehalogenation of halogenated alkanoic acids.

TCE may be degraded by the fortuitous action of enzymes involved with other catabolic pathways. For example, TCE is degraded aerobically by methanotrophs to TCE epoxide by the fortuitous action of methane monooxygenases and spontaneous abiotic transformations (Figure 9).[60,61] TCE epoxide, in spontaneous

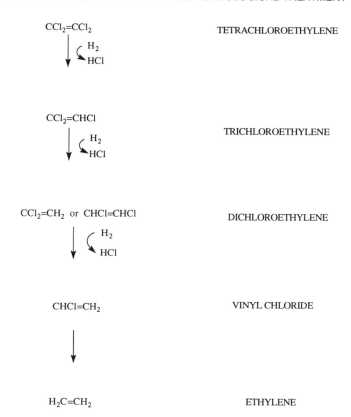

Figure 8.   Reductive transformation of tetrachloroethylene, trichloroethylene, dichloroethylene, and vinyl chloride. From Vogel and McCarty;[57] Freedman and Gossett;[58] and Hardman.[50]

reactions with water, will break down to form dichloroacetic acid, TCE-diol, formic acid, and glyoxylic acid. In addition, the chloral hydrate produced during the oxidation of TCE by methanotrophs can be oxidized by microorganisms to trichloroacetic acid or reduced to trichloroethanol.[62] Further metabolism of these intermediates results in mineralization to carbon dioxide.

TCE may also be degraded by enzymes involved in aromatic hydrocarbon degradation. Strain G4, tentatively identified as a member of the genus *Acinetobacter*, mineralizes TCE with one or more enzymes from aromatic biodegradative pathways.[63] An initial attack by a monooxygenase or a dioxygenase to form epoxide or dioxetane intermediates, respectively, was proposed. Similarly, *Alcaligenes eutrophus* strain JMP134 degrades TCE with enzymes from two independent aromatic degradation pathways: the phenol biodegradative pathway and the 2,4-dichlorophenoxyacetic acid biodegradative pathway.[64] Short chain haloalkanes are degraded by *Xanthobacter autotrophicus* GJ10 to corresponding alcohols by hydrolytic release catalyzed by a haloalkane dehalogenase.[55]

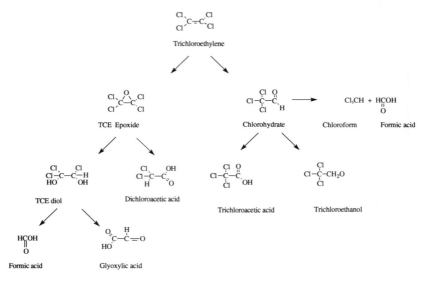

Figure 9. The degradation of trichloroethylene by aerobic bacteria and abiotic transformations. Compiled from Little et al.[61] and Newman and Wackett.[62]

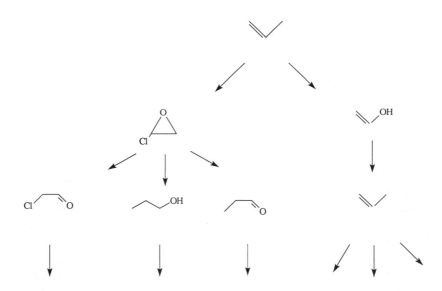

Figure 10. Aerobic degradation of vinyl chloride. Compiled from Castro et al.[68] and Castro et al.[69]

The aerobic degradation of vinyl chloride has been reported for axenic cultures and in groundwater[64-67] and degradative pathways determined (Figure 10).[68,69] Vinyl chloride is metabolized by resting cell suspensions of *Methylosinus trichosporium* OB-3b via the epoxide, chloroethylene oxide.[68] Further degradation to carbon dioxide and waters occurs by biological and chemical transforma-

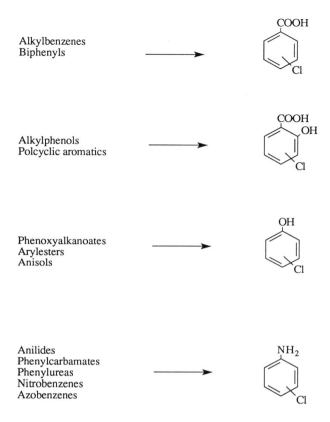

Figure 11. Generalized scheme for the degradation of chlorinated aromatic hydrocarbons. Redrawn from Reineke and Knackmuss.[71]

tions. The dominant biological pathway is the degradation of chloroethylene oxide to ethylene, then sequential degradation to ethylene glycol, hydroxyacetaldehyde, glycolic acid, and carbon dioxide. The dehalogenation of vinyl chloride by the direct hydrolysis of the C-Cl bond by a *Pseudomonas* sp. results in the formation of acetaldehyde. Further oxidation of the carboxy and methyl groups produces hydroxyacetaldehyde and acetic acid.[69] These compounds are oxidized to glycolic acid and metabolized to carbon dioxide.

4.2.2 Halogenated Aromatic and Polycyclic Aromatic Compounds

The microbial transformation of halogenated aromatics has been reported for halogenated aromatic compounds such as halogenated benzenes, benzoates, phenols, phenoxyacetates, anilines, and biphenyls (Figure 11).[70–72] Dehalogenation and degradation of halogenated aromatic compounds can be mediated by anaerobic and aerobic microbial populations and can occur prior to ring cleavage by reductive, hydrolytic, or oxygenolytic elimination.[50,71] Alternatively, dehalogenation can be spontaneously eliminated after the generation of nonaromatic structures.[71]

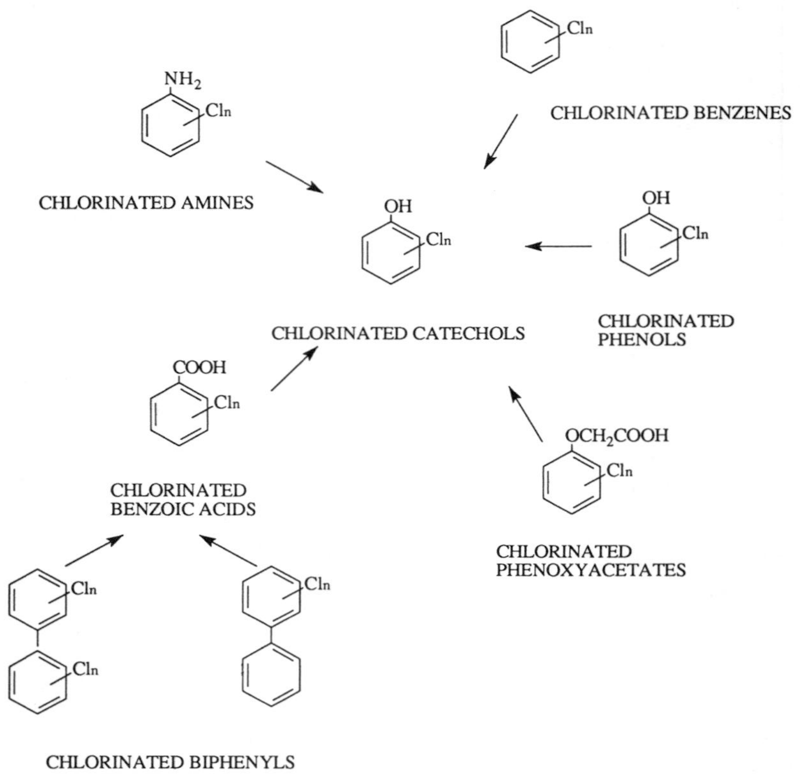

Figure 12.   Generalized scheme for the degradation of chlorinated aromatic hydrocarbons to chlorocatechols. From Reineke and Knackmuss.[71]

The degradation of halogenated aromatic compounds often proceeds via intermediates common to the catabolism of halogenated aromatics as well as the catabolic pathways for aromatic and polycyclic aromatic hydrocarbons. Some enzymes in the aromatic hydrocarbon pathways act fortuitously on the halogenated analogues. A generalized scheme for the metabolism of chlorinated aromatic compounds to chlorocatechols is shown in Figure 12. A more detailed scheme with respect to position of the halogen substituent and the degradation of chlorocatechols via ring cleavage reactions to maleacetates are depicted in Figures 13 and 14.

Reductive dehalogenation of haloaromatics has been reported for anaerobic sulfidogenic or methanogenic conditions. There is a direct substitution of a hydrogen for the halogen. Anaerobic degradation of halogenated benzoates has been detected in enrichment and pure cultures. Dehalogenation of halogenated benzoates by the isolate *Desulfomonile tiedjei* appears to be mediated by enzyme-catalyzed reactions.[72] Although unable to utilize 3-chlorobenzoate for growth, this organism is the principle dehalogenating microorganism in a defined methanogenic consortium, which mineralizes 3-chlorobenzoate. The growth of the community,

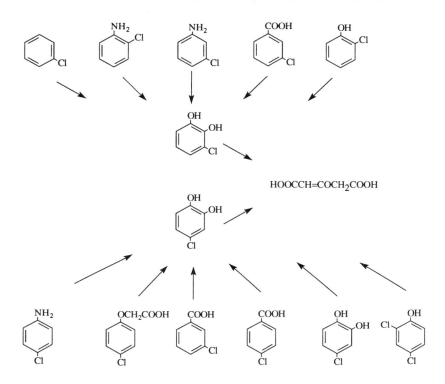

Figure 13. Degradation of chlorinated aromatic compounds to chlorocatechols and methylmaleylacetate. Compiled from Reineke and Knackmuss.[71]

as a whole, is supported by the degradation of 3-chlorobenzoate by *Desulfomonile tiedjei*.

Oxygenolytic transformation of haloaromatics is a fortuitous reaction catalyzed by dioxygenase enzyme and results in the simultaneous incorporation of both atoms of molecular oxygen as hydroxyls and the nonenzymatic loss of halogen constituent.[50-52,71,73,74] Dioxygenase activity was reported for the degradation of chlorophenylacetates, fluorinated benzenes, and polychlorinated biphenyls. Dehalogenation can also occur spontaneously after ring cleavage as in the case of the formation and cleavage of chlorocatechols. This pathway has been reported for the degradation of chlorinated benzoates, benzenes, and phenols.

Hydrolytic dehalogenation by a water-derived hydroxyl group results in the hydroxylation of the aromatic ring and the simultaneous loss of the halogen substituent.[50-52] Hydrolytic dehalogenation can occur under aerobic and anaerobic conditions. Displacement of a halogen substituent with a hydroxyl group has been demonstrated for 4-chlorobenzoate, 4-fluorobenzoate, 4-bromobenzoate, and pentachlorophenol.

Chlorinated phenols are of environmental concern because the toxicity of these compounds as well as their recalcitrance to microbial degradation increases with the degree of chlorination.[50-52] Aerobic degradation of mono- and dichlorophenols

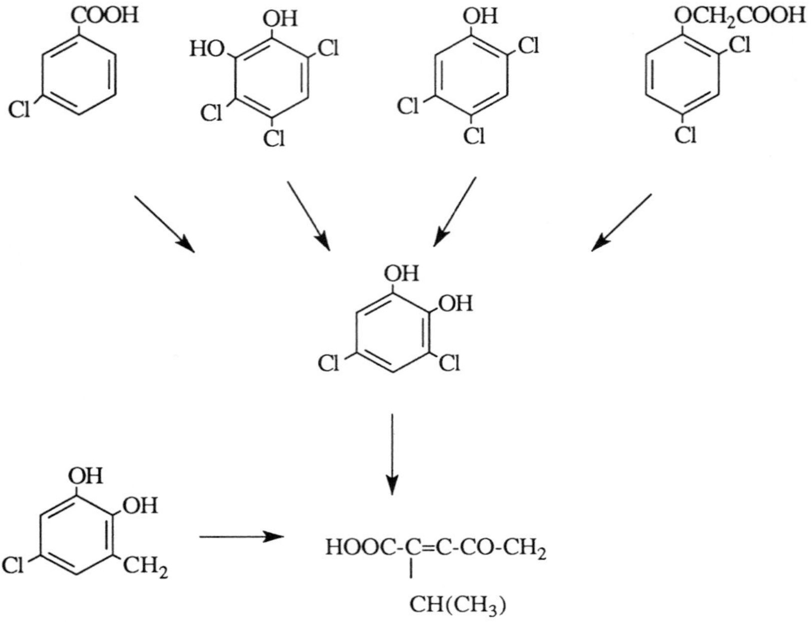

Figure 14.   Degradation of chlorinated aromatic compounds to chlorocatechols and maleylacetate. Compiled from Reineke and Knackmuss.[71]

proceeds via the formation of chlorocatechols and dechlorination following ring cleavage. Aerobic degradation of more highly chlorinated phenols proceeds via the formation of tetra-, tri-, and di-chlorohydroquinone intermediates prior to ring cleavage.[75] Pentachlorophenol (PCP) is degraded aerobically or anaerobically by a variety of bacteria and fungi, in pure and in mixed cultures. Soil isolates belonging to the genus *Flavobacterium* are able to utilize PCP for growth.[76] Anaerobic degradation of PCP proceeds via tri-, di-, and mono-dehalogenated phenolic intermediates prior to ring cleavage *Rhodococcus chlorophenolicus* PCP-1 enzyme, which fortuitously functions as a dehalogenating enzyme for PCP.[77]

   The resistance of polychlorinated biphenyls (PCBs) to microbial degradation is dependent on the number and position of the chlorines and whether the halogens are present on one or both of the aromatic rings.[78] A generalized scheme for the degradation of PCBs is shown in Figure 15.

5. BIODEGRADATION BY MIXED POPULATIONS

   The majority of information available on metabolic pathways is derived from pure cultures or defined consortia cultivated in relatively simple systems with a single compound of interest. Researchers often extrapolate information obtained from these studies to the degradation of hazardous organics in physically and

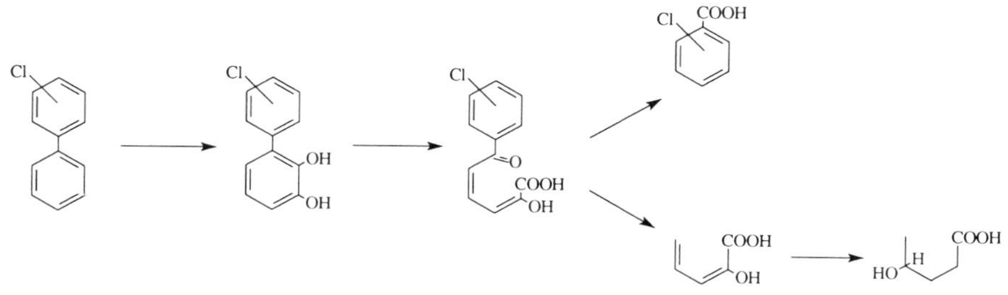

Figure 15. Degradation of polychlorinated biphenyls. Redrawn from Commandeur and Parsons.[52]

chemically heterogenous natural systems with mixed microbial populations or to bioreactors in which sterility, though often desired, is not achieved. The biochemistry of biodegradation would be, undoubtably, more complex in natural systems due to the interaction between community members, competition for limited substrates, exchange of metabolites, and variations in anaerobiosis. The biochemistry involved with the biodegradation of hazardous organics by mixed microbial communities is related to the structure of the community and the function of the community members.

An excellent illustration of the degradation of organics by mixed populations is the microbial degradation of petroleum. While petroleum is often viewed as a single substance, it is actually a complex mixture of organics consisting of linear, branched, and cyclic alkanes, polycyclic aromatic hydrocarbons, and polycyclic heteroatomic hydrocarbons.[79] Thus, the microbial degradation of petroleum is the result of the combined catabolic activities that target the different classes of organics. In fact, microorganisms with the capability of degrading aliphatic hydrocarbons are frequently ineffective against aromatic hydrocarbons[80] and *vice versa*.[81]

Through the combined activities of its members, microbial constortia may achieve mineralization of hazardous organic compounds. An excellent illustration of this is the anaerobic microbial consortium that mineralizes chlorobenzoate.[6,72] The initial step in the degradative sequence is the cometabolic conversion of 3-chlorobenzoate to benzoate by *Desulfuromonas tiedje*. This transformation, which is not directly beneficial to *D. tiedje*, is key to the support of the community as a whole.

The degradation of hazardous organics may also be feasible through the combined activities of fortuitously acting enzymes present in one or more microorganisms. However, the more complex the degradative sequence that is required, the less likely the necessary enzymes may be found in a single organism. In these cases, complete destruction may be achieved through the combined activities of fortuitously acting enzymes found in multiple microorganisms. Alternatively, the combination of complementary reactions within a single microorganism can be achieved through molecular genetics.

Examples of fortuitously acting enzymes are the action of methane monooxygenase[60,61] and the involvement of at least one enzyme in the aromatic degradation pathway in the degradation of TCE.[63] An enzyme or enzymes in the *meta* fission pathway of isolate G4, tentatively identified as a member of the genus *Acinetobacter*, are involved in the degradation of trichloroethylene.[82] The postulated pathway involves the formation of an epoxide or dioxetane by a monooxygenase or dioxygenase, respectively. Decomposition of unstable hydroxy- or oxochlorinated intermediates would result in formation of dehalogenated compounds, which would be further metabolized to CO_2 or assimilated into cell biomass.

6. CONCLUSIONS

In summary, organic compounds amenable to biologically-based hazardous waste treatments are those compounds that can be enzymatically modified. Preferred for biological waste treatment are those compounds that can be mineralized by axenic or mixed cultures. The mineralization of organics by microorganisms is a destructive process. Thus, treatment methods based on the microbial mineralization of wastes are viewed as technologies that offer permanent and cost-effective solutions to waste treatment.

7. ACKNOWLEDGMENTS

The author wishes to thank Albert J. Tien for comments on the draft of this manuscript and Kimberley Arehart for graphics. This work was supported, in part, by the Department of Energy to the Idaho National Engineering Laboratory under contract No. DE-AC07-76ID01570.

8. REFERENCES

1. Lee, M. D., and R. L. Raymond, Sr. "Case History of the Application of Hydrogen Peroxide as an Oxygen Source for *In Situ* Bioreclamation," in *In Situ Bioreclamation: Applications and Investigations for Hydrocarbon and Contaminated Site Remediation*, R. E. Hinchee and R. F. Olfenbuttel, Eds. (Boston: Butterworth-Heinemann, 1991), pp. 429–436.
2. Hutchins, S. R., and J. T. Wilson. "Laboratory and Field Studies on BTEX Biodegradation in a Fuel-Contaminated Aquifer Under Denitrifying Conditions," in *In Situ Bioreclamation: Applications and Investigations for Hydrocarbon and Contaminated Site Remediation*, R. E. Hinchee and R. F. Olfenbuttel, Eds. (Boston: Butterworth-Heinemann, 1991), pp. 157–172.
3. "Bioremediation in the Field," U. S. Environmental Protection Agency, Report EPA/540/2-91/007 No. 2. March 1991.
4. Alexander, M. L. "Biodegradation of Chemicals of Environmental Concern," *Science* 211:132–138 (1981).
5. Grbić-Galić, D., and T. M. Vogel. "Transformation of Toluene and Benzene by Mixed Methanogenic Cultures," *Appl. Environ. Microbiol.* 53:254–260 (1987).

6. Shelton, D. R., and J. M. Tiedje. "Isolation and Partial Characterization of Bacteria in an Anaerobic Consortium that Mineralizes 3-Chlorobenzoic Acid," *Appl. Environ. Microbiol.* 48:840–848 (1984).

7. Adriaens, P., and W. J. Hickey. "Physiology of Biodegradative Microorganisms," in *Biotechnology for the Treatment of Hazardous Waste*, D. L. Stoner, Ed. (Chelsea, MI: Lewis Publishers, 1993), pp. 97–136.

8. Atlas, R. M. "Microbial Degradation of Petroleum Hydrocarbons: An Environmental Perspective," *Microbiol. Rev.* 45:180–209 (1981).

9. Leahy, J. G., and R. R. Colwell. "Microbial Degradation of Hydrocarbons in the Environment," *Microbiol. Rev.* 54:305–315 (1990).

10. Atlas, R. M. *Petroleum Microbiology* (New York: Macmillan Publishing Company, 1984).

11. Bartha, R. "Biotechnology of Petroleum Pollutant Biodegradation," *Microb. Ecol.* 12:155–172 (1986).

12. Watkinson, R. J., and P. Morgan. "Physiology of Aliphatic Hydrocarbon-Degrading Microorganisms," *Biodegradation* 1:19–92 (1990).

13. Smith, M. R. "The Biodegradation of Aromatic Hydrocarbons by Bacteria," *Biodegradation* 1:191–206 (1990).

14. Kämpfer, P., M. Steiof, and W. Dott. "Microbiological Characterization of a Fuel-Oil Contaminated Site Including Numerical Identification of Heterotrophic Water and Soil Bacteria," *Microb. Ecol.* 21:227–251 (1991).

15. Hinchee, R. E., and R. F. Olfenbuttel, Eds. *In Situ Bioreclamation: Applications and Investigations for Hydrocarbon and Contaminated Site Remediation* (Boston: Butterworth-Heinemann, 1991).

16. Hinchee, R. E., and R. F. Olfenbuttel, Eds. *On Site Bioreclamation: Processes for Xenobiotic and Hydrocarbon Treatment* (Boston: Butterworth-Heinemann, 1991).

17. Bossert, I., and R. Bartha. "The Fate of Petroleum in Soil Ecosystems," in *Petroleum Microbiology*, R. M. Atlas, Ed. (New York: Macmillan Publishing Company, 1984), pp. 435–473.

18. Cooney, J. J. "The Fate of Petroleum Pollutants in Freshwater Ecosystems," in *Petroleum Microbiology*, R. M. Atlas, Ed. (New York: Macmillan Publishing Company, 1984), pp. 399–433.

19. Floodgate, G. D. "The Fate of Petroleum in Marine Ecosystems," in *Petroleum Microbiology*, R. M. Atlas, Ed. (New York: Macmillan Publishing Company, 1984), pp. 355–397.

20. Singer, M. E., and W. R. Finnerty. "Microbial Metabolism of Straight-Chain and Branched Alkanes," in *Petroleum Microbiology*, R. M. Atlas, Ed. (New York: Macmillan Publishing Company, 1984), pp. 1–59.

21. Perry, J. J. "Microbial Metabolism of Cyclic Alkanes," in *Petroleum Microbiology*, R. M. Atlas, Ed. (New York: Macmillan Publishing Company, 1984), pp. 61–97.

22. Stirling, L. A., R. J. Watkinson, and I. J. Higgins. "Microbial Metabolism of Alicyclic Hydrocarbons: Isolation and Properties of a Cyclohexane Degrading Bacterium," *J. Gen. Microbiol.* 99:119–125 (1977).

23. Magor, A. M., J. Warburton, M. K. Trower, and M. Griffin. "Comparative Study of the Ability of Three *Xanthobacter* Species to Metabolize Cycloalkanes," *Appl. Environ. Microbiol.* 52:665–671 (1986).

24. Trower, M. K., R. M. Buckland, R. Higgins, and M. Griffin. "Isolation and Characterization of a Cyclohexane-Metabolizing *Xanthobacter* sp.," *Appl. Environ. Microbiol.* 49:1282–1289 (1985).

25. Aeckersberg, M., F. Bak, and F. Widdel. "Anaerobic Oxidation of Saturated Hydrocarbons to CO_2 by a New Type of Sulfate-Reducing Bacterium," *Arch. Microbiol.* 156:5–14 (1991).

26. Dagley, S. "Biochemistry of Aromatic Hydrocarbon Degradation in Pseudomonads," in *The Bacteria: A Treatise on Structure and Function, Vol 10. The Biology of the Pseudomonas*, J. R. Soktach and L. N. Ornston, Eds. (Orlando, FL: Academic Press, Inc., 1986), Chapter 15.

27. Cerniglia, C. E. "Microbial Transformation of Aromatic Hydrocarbons," in *Petroleum Microbiology*, R. M. Atlas, Ed. (New York: Macmillan Publishing Company, 1984), pp. 99–127.

28. Stanier, R. Y., and L. N. Ornston. "The β-Ketoadipate Pathway," *Adv. Microbiol. Physiol.* 9:89–151 (1973).

29. Berry, D. F., A. J. Francis, and J.-M. Bollag. "Microbial Metabolism of Homocyclic and Heterocyclic Aromatic Compounds Under Anaerobic Conditions," *Microbiological Rev.* 51:43–59 (1987).

30. Evans, W. C., and G. Fuchs. "Anaerobic Degradation of Aromatic Compounds," *Annu. Rev. Microbiol.* 42:289–317 (1988).

31. Grbić-Galić, D. "Anaerobic Microbial Transformation of Nonoxygenated Aromatic and Alicyclic Compounds in Soil, Subsurface, and Freshwater Sediments," in *Soil Biochemistry, Vol. 6*, J.-M. Bollag and G. Stotzky, Eds. (New York: Marcel Dekker, Inc., 1990), pp. 117–189.

32. Zeyer, J., P. Eicher, J. Dolfing, and R. P. Schwarzenbach. "Anaerobic Degradation of Aromatic Hydrocarbons," in *Biotechnology and Biodegradation*, D. Kamely, A. Chakrabarty, and G. S. Omenn, Eds. (Houston, TX: Gulf Publishing Company, 1990), pp 33–47.

33. Vogel, T. M., and D. Grbić-Galić. "Incorporation of Oxygen from Water into Toluene and Benzene During Anaerobic Fermentative Transformation," *Appl. Environ. Microbiol.* 52:200–202 (1986).

34. Evans, P. J., D. T. Mang, K. S. Kim, and L. Y. Young. "Anaerobic Degradation of Toluene by a Denitrifying Bacterium," *Appl. Environ. Microbiol.* 57:1139–1145 (1991).

35. Evans, P. J., W. Ling, B. Goldschmidt, E. R. Ritter, and L. Y. Young. "Metabolites Formed During the Anaerobic Transformation of Toluene and o-Xylene and Their Proposed Relationship to the Initial Steps of Toluene Mineralization," *Appl. Environ. Microbiol.* 58:496–501 (1992).

36. Evans, P. J., D. T. Mang, and L. Y. Young. "Degradation of Toluene and m-Xylene and Transformation of o-Xylene by Denitrifying Enrichment Cultures," *Appl. Environ. Microbiol.* 57:450–454 (1991).

37. Edwards, E. A., L. E. Wills, M. Reinhard, and D. A. Grbić-Galić. "Anaerobic Degradation of Toluene and Xylene by Aquifer Microorganisms Under Sulfate-Reducing Conditions," *Appl. Environ. Microbiol.* 58:794–800 (1992).

38. Bak, F., and F. Widdel. "Anaerobic Degradation of Phenol and Phenol-Derivatives by *Desulfobacterium phenolicum* sp. nov.," *Arch. Microbiol.* 146:177–180 (1986).

39. Widdel, F., G.-W. Kohring, and F. Mayer. "Studies on Dissimilatory Sulfate-Reducing Bacteria that Decompose Fatty Acids," *Arch. Microbiol.* 134:286–294 (1983).

40. Lovley, D. R., and D. J. Lonergan. "Anaerobic Oxidation of Toluene, Phenol, and p-Cresol by the Dissimilatory Iron-Reducing Organism, GS-15," *Appl. Environ. Microbiol.* 56:1858–1864 (1990).

41. Lovley, D. R., M. J. Baedecker, D. J. Lonergan, I. M. Cozzarelli, E. J. P. Phillips, and D. I. Siegel. "Oxidation of Aromatic Contaminants Coupled to Microbial Iron Reduction," *Nature (London)* 339:297–299 (1989).

42. Grbić-Galić, D., and L. Y. Young. "Methane Fermentation of Ferulate and Benzoate: Anaerobic Degradation Pathways," *Appl. Environ. Microbiol.* 50:292–297 (1985).

43. Beller, H. R., M. Reinhard, and D. Grbić-Galić. "Metabolic By-Products of Anaerobic Toluene Degradation by Sulfate-Reducing Enrichment Cultures," *Appl. Environ. Microbiol.* 58:3192–3195 (1990).

44. Altenschmidt, U., and G. Fuchs. "Anaerobic Degradation of Toluene in Denitrifying *Pseudomonas* sp.: Indication for Toluene Methylhydroxylation and Benzoyl-CoA as Central Aromatic Intermediate," *Arch. Microbiol.* 156:152–158 (1991).

45. Bossert, I. D., and L. Y. Young. "Anaerobic Oxidation of *p*-Cresol by a Denitrifying Bacterium," *Appl. Environ. Microbiol.* 52:1117–1122 (1986).

46. Tschech, A., and G. Fuchs. "Anaerobic Degradation of Phenol by Pure Cultures of Newly Isolated Denitrifying Pseudomonads," *Arch. Microbiol.* 148:213–217 (1987).

47. Dutton, P., and W. C. Evans. "The Metabolism of Aromatic Compounds by *Rhodopseudomonas palustris*," *Biochem. J.* 113:525–536 (1969).

48. Hutber, G. N., and D. W. Ribbons. "Involvement of Coenzyme A Esters in the Metabolism of Benzoate and Cyclohexanecarboxylate by *Rhodopseudomonas palustris*," *J. Gen. Microbiol.* 129:2413–2420 (1983).

49. Williams, R. J., and W. C. Evans. "The Metabolism of Benzoate by *Moraxella* sp. Through Anaerobic Nitrate Respiration. Evidence for a Reductive Pathway," *Biochem. J.* 148:1–10 (1975).

50. Hardman, D. J. "Biotransformation of Halogenated Compounds," *Crit. Rev. Biotechnol.* 11:1–40 (1991).

51. Chaudry, G. R., and S. Chapalamadugu. "Biodegradation of Halogenated Organic Compounds," *Microbiol. Rev.* 55:59–79 (1991).

52. Commandeur, L. C. M., and J. R. Parsons. "Degradation of Halogenated Aromatic Compounds," *Biodegradation* 1:207–220, (1990).

53. Foy, C. L. "The Chlorinated Aliphatic Acids," in *Herbicides: Chemistry, Degradation, and Mode of Action, Vol 1*, P. C. Kearney and D. D. Kaufmann, Eds. (New York: Marcel Dekker, Inc., 1975), pp. 399–451.

54. Hardman, D. J., P. C. Gowland, and J. H. Slater. "Plasmids from Soil Bacteria Enriched on Halogenated Alkanoic Acids," *Appl. Environ. Microbiol.* 51:44–51 (1986).

55. Janssen, D. B., A. Scheper, L. Dijkhuizen, and B. Witholt. "Degradation of Halogenated Aliphatic Compounds by *Xanthobacter autotrophicus* GJ10," *Appl. Environ. Microbiol.* 49:673–677 (1985).

56. Fathepure, B. Z., and S. A. Boyd. "Dependence of Tetrachloroethylene Dechlorination on Methanogenic Substrate Consumption by *Methanosarcina* sp. Strain DCM," *Appl. Environ. Microbiol.* 54:2976–2980 (1988).

57. Vogel, T. M., and P. L. McCarty. "The Biotransformation of Tetrachloroethylene to Trichloroethylene, Dichloroethylene, Vinyl Chloride, and Carbon Dioxide Under Methanogenic Conditions," *Appl. Environ. Microbiol.* 49:1080–1083 (1985).

58. Freedman, D. L., and J. M. Gossett. "Biological Reductive Dechlorination of Tetrachloroethylene and Trichloroethylene Under Methanogenic Conditions," *Appl. Environ. Microbiol.* 55:2144–2151 (1989).

59. Stefano, D., J. M. Gossett, and S. H. Zinder. "Reductive Dechlorination of High Concentrations of Tetrachloroethane to Ethene by an Anaerobic Enrichment Culture in the Absence of Methanogenesis," *Appl. Environ. Microbiol.* 57:2287–2292 (1991).

60. Fogel, M. M., A. R. Taddeo, and S. Fogel. "Biodegradation of Chlorinated Ethenes by a Methane-Utilizing Mixed Culture," *Appl. Environ. Microbiol.* 54:720–724 (1986).

61. Little, C. D., A. V. Palumbo, S. E. Herbes, M. E. Lidstrum, R. L. Tyndall, and P. J. Gilmer. "Trichloroethylene Biodegradation by a Methane-Oxidizing Bacterium," *Appl. Environ. Microbiol.* 54:951–956 (1988).

62. Newman, L. M., and L. P. Wackett. "Fate of 2, 2, 2-Trichloroacetaldehyde (Chloral Hydrate) Produced During Trichloroethylene Oxidation by Methanotrophs," *Appl. Environ. Microbiol.* 57:2399–2402 (1991).

63. Nelson, M. J. K., S. O. Montgomery, W. R. Mahaffey, and P. H. Pritchard. "Biodegradation of Trichloroethylene and Involvement of an Aromatic Biodegradative Pathway," *Appl. Environ. Microbiol.* 53:949–954 (1987).

64. Harker, A. R., and Y. Kim. "Trichloroethylene Degradation by Two Independent Aromatic-Degrading Pathways in *Alcaligenes eutrophus* JMP134," *Appl. Environ. Microbiol.* 56:1179–1181 (1990).

65. Hartmans, S., J. A. M. de Bont, J. Tranper, and K. Ch. A. M. Luyben. "Bacterial Degradation of Vinyl Chloride," *Biotech. Lett.* 7:383–388 (1985).

66. Davis, J., and C. L. Carpenter. "Aerobic Biodegradation of Vinyl Chloride in Groundwater Samples," *Appl. Environ. Microbiol.* 56:3878–3880 (1990).

67. Barrio-Lage, G., F. Z. Parson, R. M. Narbartz, P. A. Lorenzo, and H. E. Archer. "Enhanced Anaerobic Degradation of Vinyl Chloride in Groundwater," *Environ. Toxicol. Chem.* 9:403–415 (1990).

68. Castro, C. E., D. M. Riebeth, and N. O. Belser. "Biodehalogenation: The Metabolism of Vinyl Chloride by *Methylosinus trichosporium* OB-3b. A Sequential Oxidative and Reductive Pathway Through Chloroethylene Oxide," *Environ. Toxicol. Chem.* 11:749–755 (1992).

69. Castro, C.E., D. M. Wade, D. M. Riebeth, E. W. Bartnicki, and N. O. Belser. "Biodehalogenation: Rapid Metabolism of Vinyl Chloride by a Soil *Pseudomonas* sp. Direct Hydrolysis of a Vinyl C-Cl Bond," *Environ. Toxicol. Chem.* 11:757–764 (1992).

70. Gibson, S. A., and J. M. Suflita. "Anaerobic Biodegradation of 2,4,5-Trichlorophenoxyacetic Acid in Samples from a Methanogenic Aquifer: Stimulation by Short-Chain Organic Acids and Alcohols," *Appl. Environ. Microbiol.* 56:1825–1832 (1990).

71. Reineke, W., and H.-J. Knackmuss. "Microbial Degradation of Haloaromatics," *Ann. Rev. Microbiol.* 42:263–287 (1988).

72. DeWeerd, K. A., and J. M. Suflita. "Anaerobic Aryl Reductive Dehalogenation of Halobenzoates by Cell Extracts of *Desulfomonile tiedjei*," *Appl. Environ. Microbiol.* 56:2999–3005 (1990).

73. Klages, U., A. Markus, and F. Lingens. "Degradation of 4-Chlorophenylacetic Acid by a *Pseudomonas* Species," *J. Bacteriol.* 146:64–68 (1981).

74. Renganathan, V. "Possible Involvement of Toluene-2,3-Dioxygenase in Defluorination of 3-Fluoro-Substituted Benzenes by Toluene-Degrading *Pseudomonas* sp. Strain T-12," *Appl. Environ. Microbiol.* 55:330–334 (1989).

75. Steiert, J. G., and R. L. Crawford. "Microbial Degradation of Chlorinated Phenols," *Trends Biotechnol.* 3:300–305 (1985).

76. Saber, D. L., and R. L. Crawford. "Isolation and Characterization of *Flavobacterium* Strains that Degrade Pentachlorophenol," *Appl. Environ. Microbiol.* 50:1512–1518 (1985).

77. Apajalahti, J. H. A., and M. S. Salkinoja-Salonen. "Dechlorination and *para*-Hydroxylation of Polychlorinated Phenols by *Rhodococcus chlorophenolicus*," *J. Bacteriol.* 169:675–681 (1987).

78. Furukawa, K., K. Tonomura, and A. Kamibayashi. "Effect of Chlorine Substitution on the Biodegradability of Various Polychlorinated Biphenyls," *Appl. Environ. Microbiol.* 35:223–227 (1979).

79. Altgelt, K. H., and T. H. Gouw, Eds. *Chromatography in Petroleum Analysis* (New York: Marcel Dekker, Inc., 1979).

80. Fedorak, P. M., J. D. Payzant, D. S. Montgomery, and D. W. S. Westlake. "Microbial Degradation of n-Alkyl Tetrahydrothiophenes Found in Petroleum," *Appl. Environ. Microbiol.* 54:1243–1248 (1988).

81. Foght, J. M., and D. W. S. Westlake. "Degradation of Polycyclic Aromatic Hydrocarbons and Aromatic Heterocycles by a *Pseudomonas* Species," *Can. J. Microbiol.* 34:1135–1139 (1988).

82. Nelson, M. J. K., S. O. Montgomery, E. J. O'Neill, and P. H. Pritchard. "Aerobic Metabolism of Trichloroethylene by a Bacterial Isolate," *Appl. Environ. Microbiol.* 52:383–384 (1986).

83. Geissler, J. F., C. S. Harwood, and J. Gibson. "Purification and Properties of Benzoate-Coenzyme A Ligase, a *Rhodopseudomonas palustris* Enzyme Involved in the Anaerobic Degradation of Benzoate," *J. Bact.* 170:1709–1714 (1988).

CHAPTER 2

Inorganic Hazardous Waste Amenable to Biological Transformation

Henry L. Ehrlich

1. INTRODUCTION

Hundreds of millions of tons of waste are generated annually by U.S. industrial[1] and domestic activities, a significant portion of which contains inorganics that, in sufficient concentration, are toxic to human beings and other life forms. Inorganics with toxic potential include compounds of antimony, arsenic, boron, chromium, copper, cyanide, lead, mercury, nickel, selenium, silver, tin, tungsten, and uranium. An inorganic toxicant may be cationic such as metallic ions of copper, mercury, or tin or anionic such as the oxyanions chromate, molybdate, or tungstate. Toxic inorganics may also be alkylated or aromatized forms of metal ions such as methylmercury, butyltin, and phenylmercury. The inorganics may be major constituents of a waste or be present in small amounts. The toxic inorganics may be in solution form in a liquid waste or present in an insoluble form in solid waste or slurries.

0-87371-613-2/94/$0.00+$.50
© 1994 by Lewis Publishers

When improperly disposed of, insoluble toxic inorganics present in solid waste may be remobilizable through solubilization, by simple dissolution on contact with water, or by chemical reactions that may be greatly facilitated by microbes. Such mobilization results in pollution at the waste disposal site. However, the microbial transformations that mediate the mobilization of toxic inorganics may be used to treat the solid wastes and render them nontoxic prior to disposal. Microbial transformations may also be used to immobilize or concentrate the toxic inorganics in liquid wastes to render the waste safe for disposal.

This chapter will begin with a general discussion of inorganic chemical toxicity and its impact on organisms and the environment. It will continue with a discussion of the enzymatic and nonenzymatic transformations that may be used to detoxify or modify inorganic materials and, by the way of example, describe some of the applications of microbial transformation to the detoxification and treatment of inorganic hazardous waste. The chapter will close with a discussion of some potential applications of microbial transformation processes to the treatment of inorganic hazardous waste.

2. TOXICITY OF INORGANIC CHEMICALS

The toxicity of hazardous inorganics, manifested in aqueous solution, may be expressed through acute poisoning, teratogenicity, mutagenicity, or carcinogenicity. A metal toxicant may bring about one or more of these effects. If more than one is exhibited, then the effects are likely dose-dependent. Metal toxicity may affect all forms of life including microorganisms, plants, and animals, but the degree of toxicity varies for different organisms. Table 1 lists some toxic inorganics and their minimum toxic dose.

Even when the concentration of an inorganic substance with poisonous potential may be below what is considered to be toxic, the presence of this substance may, nevertheless, constitute a health hazard. This health hazard is due to bioaccumulation. A toxic inorganic may be accumulated and concentrated within a single organism by ingestion over time or concentrated through successive trophic levels in a food chain until toxic concentrations are achieved within the end member of the food chain. These cumulative effects are dependent on the fixation of the toxic substance within specific tissues of an organism or in the whole organism and on the absence of significant excretion at any trophic level. A general discussion of metal toxicity may be found in References 2 and 3.

Precise quantification of toxicity of inorganic ion species is difficult. In the case of microbial toxicity, for instance, measured toxicity depends on test conditions such as pH, redox potential, and composition of the test medium. The pH and redox potential can affect the chemical state of the toxic metal species. Complexation reactions may occur between the toxic metal species and organic constituents of the test medium. The effects of test conditions were recognized as long ago as 1949 when Schade[4] indicated that medium composition and pH affect the toxicity of metal ions for microorganisms. The effects of test conditions have been

Table 1.   Inorganic Toxicants and LD50 Doses

Compound	LD50 Dose[a]	Animal[a]
Antimony pentasulfide	1.5 g/kg	Rat
Antimony pentoxide	4.0 g/kg	Rat
Arsenic trioxide	138 mg/kg	Rat (orally)
Arsine	3.0 mg/kg	Mouse (i.p.)
Cupric sulfate	300 mg/kg	Rat (orally)
Lead dioxide	200 mg/kg	Guinea pig (i.p.)
Mercuric chloride	37 mg/kg	Rat (orally)
Nickel chloride	60 mg/kg	Dog (LD, i.v.)
Sodium chromate	243 mg/kg	Rabbit (s.c.)
Sodium cyanide	2.2 mg/kg	Rabbit (MLD, s.c.)
Sodium selenate	4 mg/kg	Rabbit (LD100 orally)
Stannous chloride	35 mg/kg	Dog (LD, i.v.)

Source: Oehme.[3]

[a]   Abbreviations: LD, lethal dose; MLD, minimum lethal dose; i.p., intraperitoneal; i.v., intravenous; s.c., subcutaneous.

repeatedly emphasized in subsequent decades as, for instance, by Sadler and Trudinger[5] and Callender and Barford.[6]

The effect of culture medium composition on the evaluation of silver ion toxicity for *Thiobacillus ferrooxidans* may be seen in the following example. Norris and Kelly[7] found that Ag^+ in a mineral salts medium with ferrous sulfate as the energy source and without added chloride was inhibitory to growth of a silver-sensitive strain of *T. ferrooxidans* at concentrations in excess of 10^{-8} M and to a silver-resistant strain (Ag-R) derived from the wild-type at concentrations in excess of 10^{-7} M. In contrast, Ehrlich[8] and Ehrlich (unpublished results) found that in the iron medium of Silverman and Lundgren,[9] which includes 1.34 mM potassium chloride among its ingredients, growth of *T. ferrooxidans* strain 19759 occurred at 10^{-4} M. Hughes and Poole[10] have attributed the greater tolerance of 10^{-4} M Ag^+ by strain 19759 to the almost complete precipitation of Ag^+ by the 1.34 mM potassium chloride present in the medium. The precipitation of Ag^+ was predicted from the solubility product for AgCl, which they cited as $10^{-9.96}$ M^2. However, this explanation is not tenable because the chloride ion was present in greater than 10-fold excess with respect to Ag^+. At 1.34 mM Cl^- concentration, Ag^+ at 10^{-1} mM is not precipitated. Given an association constant of $10^{5.24}$ (see Reference 11), it can be calculated that enough of the Ag^+ forms a soluble chloride complex ($AgCl_2^-$) to lower the effective Ag^+ concentration to a level where it is insufficient to form AgCl with the residual Cl^-.

3. PHYSIOLOGICAL BASIS OF MICROBIAL DETOXIFICATION OF HAZARDOUS WASTE CONTAINING INORGANICS

At sublethal doses, many inorganics may be detoxified or transformed through special metabolic processes. In higher animals, detoxification may be mediated by the microsomal system. For microorganisms, detoxification may be the result of specific detoxification mechanisms, distinct metabolic processes used to satisfy energy or nutrient requirements, or general microbial metabolic activities or physiological attributes with which the microorganisms happen to be endowed. Microbiological detoxification processes fall in either of two broad categories: enzymatic processes and nonenzymatic processes. Toxic inorganics may also be viewed as being modified or transformed directly or indirectly by the action of the microorganisms.

Examples of enzymatic processes include: oxidations and reductions in which the toxicant is the electron donor and electron acceptor, respectively; the production or degradation of ligands that complex toxicants; the production of lixiviants for the extraction of toxic metal compounds from solids; the volatilization of metals by transformation to organo-metals; and the production of precipitants. An example of a nonenzymatic microbiological detoxification process is the passive biosorption of toxicants from liquid wastes. The microbial generation of an environmental pH and/or redox condition that favors precipitation of an inorganic toxicant may be invoked as an indirect mechanism for the removal of toxic inorganic species from solution. However, such conditions are merely a reflection of the metabolic activity of microorganisms and may be assigned to one of the more specific enzymatic categories listed above.

3.1 Enzymatic Oxidations

In the case of inorganic species that can exist in more than one oxidation state and in which the higher oxidation state is significantly less soluble under appropriate conditions, enzymatic oxidation may be a useful way for removing the inorganic species from solution. The oxidation of Mn^{2+} to $Mn(III)$ or $Mn(IV)$ oxide is an example.[12] The oxidized product may or may not accumulate on the surface of the cells that produce it. In some instances, the cells catalyzing the oxidation can conserve energy from the oxidation, in others not. The oxidation of Fe^{2+} to $Fe(III)$ oxide or basic iron sulfate is another example.[12] Under fully oxidizing conditions at a pH below 3, Fe^{2+} oxidation proceeds rapidly only through biological catalysis by any one of a group of acidophilic iron-oxidizing bacteria such as *T. ferrooxidans*, *Leptospirillum ferrooxidans*, *Acidianus brierleyi*, *Sulfolobus acidocaldarius*, and *Sulfobacillus thermosulfidooxidans*. The first two are mesophiles with optimum growth temperatures of ~ 25°C, and the last two are thermophiles with temperature optima >45°C. All conserve energy from the oxidation of the ferrous iron. Above pH 2 but below pH 3, at least a portion of the oxidized iron will precipitate as basic ferric sulfate. At a pH well above 3, the ferric iron will precipitate mostly as ferric oxide.[13,14]

Under fully oxidizing conditions at a pH above 5, $Fe(II)$ autoxidizes rapidly and precipitates as a hydroxide or oxide; no microbial catalysis is needed or possible. On the other hand, under partially reduced conditions (low oxygen tension), microbial catalysis benefits $Fe(II)$ oxidation at a pH around neutrality.

Gallionella ferruginea oxidizes the iron and accumulates it in its stalk under these conditions. It appears to be able to conserve some of the energy liberated in the oxidation of the ferrous iron.[15,16]

In the case of inorganic species that can exist in more than one oxidation state and in which the higher oxidation state is significantly more soluble under appropriate conditions, enzymatic oxidation may be a useful way for removing the inorganic species from a solid waste. The enzymatic oxidation of metallic sulfides is an example where oxidation results in the solubilization of a solid material. For instance, the oxidation of the copper and sulfide moieties of chalcocite (Cu_2S) results in the release of Cu ions into solution.[17] The simultaneous oxidation of the metal and sulfide moieties in the case of chalcopyrite ($CuFeS_2$) is another example.[18]

3.2 Enzymatic Reduction

In the case of toxic inorganic species that can exist in more than one oxidation state and whose reduced state is insoluble under appropriate conditions, enzymatic reduction may be useful in removing the species from solution. The reduction may require anaerobic conditions (absence of oxygen in air). Anaerobic conditions in the absence of bacteria with appropriate reducing ability are usually insufficient to cause reduction of the reducible inorganic species. An example in which anaerobic conditions are essential for bacterial reduction is that of the conversion of selenate or selenite to elemental selenium, which is insoluble.[19] Another is the reduction of UO_2^{2+} to UO_2 by strain GS-15 (see Reference 20). On the other hand, CrO_4^{2-} may be bacterially reduced to Cr(III) oxide under aerobic[21] or anaerobic conditions[22,23] with a suitable electron donor such as glucose. Cr(III) oxide is sparingly soluble in water around neutral pH.

Enzymatic reduction of inorganic ions used as terminal electron acceptors during bacterial respiration may alter the solubility of toxic inorganic solids. For example, enzymatic reduction of sulfide precipitated wastes may be useful in dissolution of toxic inorganics. The activity of sulfate-reducing bacteria in $(Ba,Ra)SO_4$ sludges derived from uranium mining operations can result in the reduction of sulfate to sulfide with the simultaneous loss of barium and radium.[24]

3.3 Complexation

The formation of soluble inorganic or organic complexes of metal species can be an important mechanism of detoxification or, at least, lessening of the toxicity of the metal species. Microbes can be the source of specific complexing agents such as oxalate formed by fungi, 2-ketogluconate and various amino acids formed by bacteria, or siderophores and metallothioneins formed by bacteria and fungi. The use of complexation agents may be useful in mobilizing toxic inorganics to facilitate their removal from a solid waste in much the same manner that citric acid, organic chelators, and large heteropolymers such as fulvic acid formed microbiologically from lignin are very important complexing agents of toxic metals in soils or sediments.[25-27] Metabolic acids can also be used to dissociate insoluble metal phosphate salts by complexing with the cationic portion of the complexes.[12]

3.4 Ligand Degradation

The biodegradation of organic chelating agents may be means by which the solubility and mobility of toxic inorganic materials may be modified.[28-30] Because chelator-metal complexes are formed with coordinate ionic bonds, these associations may be disrupted by altering the pH or by oxidation/reduction reactions. For example, the reduction of Fe(III) to Fe(II) resulted in the release of iron from hydroxamate siderophore.[31]

Covalently complexed metal ions are much more toxic than the free ion and are also more difficult to disrupt than ionically associated complexes. Two typical examples are methylmercury, $[CH_3]Hg^+$, and phenylmercury, $[C_6H_5]Hg^+$, which because they have both lipophilic and hydrophilic properties, have an affinity for lipid portions of a cell and in vertebrates may show special affinity for the nervous system. Bacterial demethylation or dephenylation by mercuric lyase followed by reduction of the resultant Hg^{2+} to volatile Hg^o by mercuric reductase (see also section below) is an effective way of detoxification of the organo-mercury compounds. These transformations occur in environments polluted with these substances.[32]

3.5 Production of Precipitants and Other Forms of Metal Immobilization

Solutions containing toxic metals can be detoxified by precipitation of the metals with appropriate microbial metabolic products. Precipitation essentially removes the metals from solution. Sulfide produced anaerobically in bacterial sulfate reduction is a prime example of a metabolic product that combines with and precipitates metallic ions in solution. Most metal sulfides are quite insoluble in aqueous solution.[12,33] Some typical solubility products of metal sulfides are listed in Table 2. The stability of these sulfides depends on the maintenance of anoxic conditions. In air, many of the metal sulfides tend to autooxidize to sulfates at significant rates. Most of the sulfates tend to be very soluble in water, ferric sulfate and lead sulfate being exceptions.

Microbial degradation of organo-phosphates with the release of orthophosphate can lead to metal precipitation through the formation of metal phosphates, especially at pH values at and above neutrality. The precipitate may be associated with the cells responsible for generating the orthophosphate.[34] This process can be practically exploited by use of immobilized cells.[35-37]

In other instances, metal phosphate precipitates may form external to cells that generate orthophosphate from phosphate esters using extracellular phosphatases.[38-40] Once formed, however, these phosphates may at least partially redissolve if the

Table 2.   Solubility Products of Some Typical Metal Sulfides

Compound	Solubility Product[a]
Cadmium sulfide (CdS)	$(10^{-28.4})$
Cobalt sulfide (CoS)	$(10^{-25.5})$
Cuprous sulfide (Cu2S)	$(10^{-46.7})$
Cupric sulfide (CuS)	$(10^{-44.1})$
Ferrous sulfide (FeS)	$(10^{-18.4})$
Lead sulfide (PbS)	$(10^{-27.5})$
Mercuric sulfide (HgS)	$(10^{-52.4}$ to $10^{-48.7})$
Nickel sulfide (NiS)	$(10^{-23.9})$
Zinc sulfide (ZnS)	$(10^{-22.9})$

Source: Weast and Astle.[89]

[a]   All constants determined at 18°C.

environmental pH falls below neutrality. Intracellular polyphosphates may also immobilize metals taken into cells. An example of this process is the immobilization of aluminum by the cyanobacterium *Anaboena cylindrica*.[41]

3.6 Biosorption

Biosorption of toxic inorganics may be mediated by nonenzymatic processes such as adsorption or flocculation, while bioaccumulation tends to be mediated by enzymatic processes such as intracellular transport and accumulation. Biosorption can exhibit characteristics of complexation, coordination, ion exchange, and inorganic precipitation.[42] Adsorption and flocculation are usually due to nonspecific binding of ionic species to cell surface-associated or extracellular polysaccharide and proteins, or particulate matter. Biosorption is not dependent on cell metabolism. Active transport and intracellular accumulation are typically enzymatically-mediated energy requiring processes that necessitate live biomass.

Metallic ions can also be removed from solution by sorption to particulates, which include living and dead biomass as well as organic debris from the breakdown of plant and animal matter.[43–45] This metal immobilization may be reversed, however, through displacement by ion exchange including proton exchange when an adsorption mechanism is involved. The tendency for such exchange to occur depends on the relative affinity of the adsorbent for the adsorbed metal and the potential exchanging ion species.

It is clear that *in situ* metal immobilization by any of the mechanisms mentioned in this section may not always result in a permanent detoxification of a metal-polluted environment.

3.7 Volatilization by the Formation of Organo-Complexes

Some metal species polluting an environment may be removed by volatilization. In the case of mercury, this frequently involves the bacterial reduction of mercuric ion to volatile metallic mercury.[12,32] This process depends on the special property of mercury in the metallic state that keeps it liquid with a high vapor pressure at ambient temperature and atmospheric pressure. Other metals and metalloids, like cadmium, lead, tin, arsenic, selenium, and tellurium, have to be methylated to be volatilized. Even mercuric mercury can be volatilized as dimethylmercury, but the formation of this compound involves a very inefficient mechanism. The methylation of metals seems to occur mostly as a result of nonenzymatic interaction with a microbially formed methyl donor such as methylcobalamine in the case of mercury and a methyl halide such as methyl iodide in the case of other metals.[46-51] Methylation of metalloids, on the other hand, is directly linked to reductive metabolism of these compounds.[52]

4. HAZARDOUS INORGANIC WASTE AMENABLE TO MICROBIAL TREATMENT

Solid and/or liquid wastes containing toxic inorganics may be generated in various industrial processes such as chemical manufacturing, electric power generation, coal and ore mining, hydrometallurgical ore processing, smelting and metal refining, metal plating, tanning, and others. Sludge generated in treatment of domestic waste may also contain undesirable levels of toxic inorganics. Inorganics in solid waste, if they are in a chemically and biologically unreactive form, (i.e., they cannot be mobilized and enter the soil and/or an aquifer in significant quantities or at a significant rate) may be considered to present no toxicity problem when disposed of in the environment. On the other hand, when inorganics are in a reactive form allowing for their mobilization, they can enter the soil and/or aquifer.[53] Therefore, these inorganics must be removed from the waste before they can be disposed of in the environment. Such removal may be accomplished by physicochemical or microbiological treatment involving some kind of leaching process. The leaching may consist of an aqueous extraction if the inorganic substance is a water-soluble compound. However, aqueous extraction may be enhanced through use of complexing agents or acidulants that facilitate metal solubilization. These complexing agents and acidulants may be economically produced by appropriate microbes. If the inorganic toxicant in a solid waste is in an insoluble but reactive form, it may require oxidation or reduction to render the compound soluble before it can be extracted. The oxidation or reduction in a controlled process may be purely chemical, or it may be mediated through microbial action. The extraction of the inorganics from the solid waste constitutes detoxification of that waste in respect to these substances.

Liquid waste containing inorganics at toxic levels must be detoxified by removal of the inorganics before it can be safely disposed of in the environment. As in the case of solid waste, this removal may be accomplished physicochemically or microbiologically in a controlled process. Regardless of whether a physicochemical or microbiological treatment is used, the removal process may involve sorption or precipitation. Sorption by chemical binding (e.g., ion exchange) and/ or physical binding (e.g., by Van der Waals forces) can be accomplished by commercially available ion exchangers, adsorbers such a charcoal and diatomaceous earth, as well as commercially available biosorbents.[45] Nonbiological precipitation may be accomplished by pH adjustment to a range where the dissolved toxicant becomes insoluble or sparingly soluble. Nonbiological precipitation may also be accomplished by the addition of a chemical reagent that reacts with the dissolved toxic species to form an insoluble compound. The liming of acid mine drainage to raise its pH to a range at which iron precipitates as iron oxide or basic ferric sulfate and coprecipitates other toxic metal ions is an example of toxicant removal by nonbiological pH adjustment. The reduction of chromate in tanner's waste to insoluble chromium oxide by reaction with a ferrous salt is an example of toxicant removal by nonbiological, chemical reduction. Biological precipitation may involve the microbial generation of a compound that upon reacting with the dissolved toxicant precipitates it. The anaerobic generation of H_2S from sulfate by sulfate-reducing bacteria is an example. Many base metal ions form very insoluble sulfides. Other forms of biological precipitation of a dissolved toxic metal species may involve direct oxidation or reduction, in which case the oxidized or reduced product is an insoluble compound. Bacterial oxidation of Mn^{2+} to manganese oxide[12] or the reduction of UO_2^{2+} to UO_2 (see Reference 20) would be examples of such reactions.

5. EXAMPLES OF BIODETOXIFICATION PROCESS APPLICATIONS

5.1 Iron Removal from Industrial Waste Streams

In Japan, the ability of *Thiobacillus ferrooxidans* to oxidize ferrous to ferric iron has been exploited in the removal of iron from acid mine drainage at the Yanahara pyrite mine and the abandoned Matsuo sulfur-pyrite mine. The organisms in this process are immobilized in a fixed film on diatomaceous earth and used in a reactor with forced aeration. Complete iron and arsenic precipitation after bacterial oxidation is ensured by reacting the treatment effluent with calcium carbonate after separation and recycling of the bacterially coated diatomaceous earth.[54,55] At the Kosaka smelter of the Dowa Mining Co., *Thiobacillus ferrooxidans* immobilized on diatomaceous earth are used to remove the iron left in a leachate that resulted when flue dust was extracted with sulfuric acid and from which copper, iron, and zinc were first recovered. In this process, the bacterially coated diatomaceous earth is also recycled.[55]

At Barite Industries Co. in Kosaka, Japan, hydrogen sulfide waste gas is treated with ferric sulfate to produce ferrous sulfate and elemental sulfur. After recovery of the sulfur, the ferrous sulfate is oxidized to ferric sulfate with the help of *Thiobacillus ferrooxidans* and recycled for further oxidation of hydrogen sulfide waste gas.[55]

5.2 Cyanide Removal from Precious Metal Processing Waste Streams

A biotreatment process has been developed at the Homestake Mining Co. in Lead, South Dakota, for the degradation of cyanide in the waste stream from their gold-recovery process. This process involves the bacterial conversion of cyanide to ammonia and carbon dioxide and the subsequent bacterial oxidation of the ammonia to nitrate. The nitrate is then safely discharged into a receiving stream. The bacteria active in the process are immobilized on rotating biological contactors.[56–58]

5.3 Precious and Base Metal Recovery from Waste Streams by Biosorption

Bio-recovery Systems, Inc. (Santa Cruz, New Mexico) has successfully marketed biomass preparations of cyanobacteria (*Cyanidium* and *Spirulina*) and algal biomass (*Chlorella*) for the removal and recovery of precious metals from wastewaters by biosorption. This process is also applicable for the recovery of base metals.[59,60] A process (BIOCLAIM) based on similar principles but using processed biomass from nonphotosynthetic bacteria has been successfully applied in the removal and recovery of precious metals and the removal of low levels of toxic base metals from waste streams.[61] The biomass in all these processes is nonliving.

Successful removal of heavy metals, especially lead, from mine mill wastewater by algal growth was achieved by construction of a shallow meandering stream system in the Missouri's New Lead Belt in which the algae developed.[62,63] Pilot studies are currently under way on the use of constructed wetlands for the treatment of acid mine drainage from bituminous coal mines. Success of this process depends in part on biosorption of inorganic pollutants by living plant biomass (e.g., *Sphagnum* moss and *Typha*, commonly known as cattails) and in part on concentration by algae and bacteria through cellular uptake or catalysis of oxidations or reductions involving the inorganic pollutants.[64,65]

6. PROCESSES NOT YET APPLIED THAT HAVE BIODETOXIFICATION POTENTIALS

6.1 Radionuclide Removal from Waste Streams

Biotechnology in development that shows promise in the removal of low-level concentrations of radionuclides in waste streams involves mainly biosorption using fungal biomass,[66,67] but in the case of uranium, also the anaerobic enzymatic reduction of soluble U(VI) as uranyl ion to insoluble U(IV) by *Geobacter melallireduens* GS-15.[20]

6.2 Chromate Removal from Waste Streams

The ability of some bacteria to reduce Cr(VI) (chromate and dichromate) to Cr(III) in a respiratory process can be exploited for the removal of low-level chromate concentrations from waste streams. At neutral to alkaline pH, Cr(III) tends to form insoluble oxide-hydroxide in aqueous solution. Moreover, Cr^{3+} is much less toxic to cells than Cr(VI). Most processes described in the literature to date involve facultative organisms that reduce Cr(VI) only anaerobically,[22,23,68,69] but at least one organism, *Pseudomonas fluorescens* strain LB 300 has been found to catalyze the reduction aerobically and anaerobically.[69] All active organisms have been found to be significantly more chromate-resistant than organisms sensitive to chromate. In *P. fluorescens*, chromate resistance and the ability to reduce Cr(VI) are two distinct genetic traits.[21,70] Claims have been made for the use of chromate-reducing bacteria in sewage treatment processes to remove low levels of chromate[12,70] and in treatment of low-level chromate-polluted wastewater, in the latter case using immobilized bacteria.[70]

6.3 Selenium Removal from Surface and Groundwater

A possibility of removing selenate and selenite as pollutants of surface or groundwater microbiologically exists. The removal may be either by immobilization through anaerobic reduction of the pollutants to elemental selenium by an, as yet, unidentified bacterium[19,71] or by volatilization by a fungus such as *Alternaria alternata* through reduction and methylation to dimethylselenide.[72] The bacterial reduction process is anaerobic, whereas the fungal process is aerobic.

Immobilized, elemental selenium will be stable only as long as anaerobic conditions prevail. Under aerobic conditions, Se° can be reoxidized by certain bacteria to selenite or selenate.[12] The formation of dimethylselenide has the advantage of removing the selenium from a polluted environment rather than merely immobilizing the selenium in it. A bench-scale, plug-flow reactor has been described in which a consortium of facultative bacteria removed up to 96% of selenium from agricultural drainage waters over a period of 1 year.[73] The selenate was reduced to elemental selenium in this case.

6.4 Removal of Base Metals such as Cd, Pb, Zn, Ni, and Co from Liquid Waste by Biosorption

Biotechnologically the best method for removing cadmium, lead, zinc, nickel, cobalt, and other base metals from dilute liquid waste streams appears to be biosorption. Nonliving bacterial, fungal, and algal biomass have been shown to be effective under different conditions.[45] This technology is on the verge of finding widespread application in some industrial activities.

In situ immobilization of these metals as sulfides by bacterial sulfate reduction under anaerobic conditions can be a viable method for removing dilute concentrations of the metals from waste streams. However, as previously indicated, many of these metal sulfides may be subject to reoxidation to water soluble sulfate salts if the anaerobic environment is replaced by aerobic conditions.

6.5 Removal of Base Metals such as Cd, Pb, Zn, Ni, and Co from Solid Waste by Bioleaching

Solid wastes containing base metals that may slowly separate from it and pollute surface and/or groundwater can be leached from the wastes with the help of bacteria or fungi. In the majority of cases, the leaching is by an indirect process involving microbially produced acid and/or ligands by autotrophic or heterotrophic bacteria. The leached metal may be recovered by biosorption, as described above. Some promising approaches have been described by Bosecker.[74,75] The possibility of recovering chromium from sulfide matte prepared from superalloy scrap by leaching with sulfuric acid produced by *Thiobacillus thiooxidans* from the oxidation of elemental sulfur has been suggested by findings of Ehrlich.[76] Partially reduced sulfur species in addition to sulfuric acid generated by *T. thiooxidans* may also have played a role in this leaching process.

6.6 Pyrite Removal from Coal

Much progress has been made in demonstrating the feasibility of removing pyrite from coal by treatment with *Thiobacillus ferrooxidans*, *Acidianus*, or *Sulfolobus*. In the most widely studied process, the bacteria oxidize the pyrite to ferric sulfate, which is then separated from the coal by washing it. For efficient bacterial oxidation of the pyrite, the coal is pulverized. The progress in the development of this technology can be seen by examining the publications of Huber et al., Mishra et al., Bos et al., Beyer et al., Angus et al., and Merrettig et al.[77–83] Some success has also been reported in using bacteria such as *Thiobacillus ferrooxidans* as flotation suppressors in separating pyrite from pulverized coal.[84–86]

Economic analyses of bacterial coal desulfurization have been presented by Dugan[87] and Bos et al.[79,88] It is seen that acceptable economics depend on the quality of the raw coal to be treated, the extent of desulfurization that is required to meet environmental requirements, and process design.

7. CONCLUSION

This brief overview shows that microbes can detoxify a range of inorganic pollutants by forming less toxic compounds that persist in the affected environment, by removing them from the affected environment by making them insoluble, or by volatilizing them. Up to now, this microbial potential has been exploited on a large scale to only a very limited extent. Much wider application is possible and should be more seriously considered than it has been up to now. This will require the extensive collaboration of microbiologists, as well as biochemical and environmental engineers. It will also require persuasion through education of industrial managers, chemical and civil engineers, and government officials who are responsible for pollution control, and not least the general public that those processes that can be shown to be economic and safe are acceptable.

8. REFERENCES

1. Freeman, H. M., Ed. *Standard Handbook of Hazardous Waste Treatment and Disposal* (New York: McGraw-Hill, 1988).
2. Somani, S. M., and F. L. Cavender, Eds. *Environmental Toxicology. Principles and Policies* (Springfield, IL: Charles C. Thomas, 1981).
3. Oehme, F. W., Ed. *Toxicity of Heavy Metals in the Environment, Vols. 1 and 2* (New York: Marcel Dekker, 1978).
4. Schade, A. L. "Cobalt and Bacterial Growth, with Special Reference to *Proteus vulgaris,*" *J. Bacteriol.* 58:811–822 (1949).
5. Sadler, W. R., and P. A. Trudinger. "The Inhibition of Microorganisms by Heavy Metals," *Min. Deposita.* 2:158–168 (1967).
6. Callender, L. J., and J. P. Barford. "Precipitation, Chelation, and the Availability of Metals as Nutrients in Anaerobic Digestion. II. Applications," *Biotech. Bioeng.* 25:1959–1972 (1983).
7. Norris, P. R., and D. P. Kelly, "Toxic Metals in Leaching Systems," in *Metallurgical Applications of Bacterial Leaching and Related Microbiological Phenomena,* L. E. Murr, A. E. Torma, and J. A. Brierley, Eds. (New York: Academic Press, 1978), pp. 83–102.
8. Ehrlich, H. L. "Bacterial Leaching of Silver from a Silver-Containing Mixed Sulfide Ore by a Continuous Process," in *Fundamental and Applied Biohydrometallurgy,* R. W. Lawrence, R. M. R. Branion, and H. G. Ebner, Eds. (Amsterdam: Elsevier, 1968), pp. 77–88.
9. Silverman, M. P., and D. G. Lundgren. "Studies on the Chemoautotrophic Iron Bacterium *Ferrobacillus ferrooxidans.* I. An Improved Medium and a Harvesting Procedure for Securing High Cell Yields," *J. Bacteriol.* 77:642–647 (1959).
10. Hughes, M. N., and R. K. Poole. "Metal Speciation and Microbial Growth — the Hard (and Soft) Facts," *J. Gen. Microbiol.* 137:725–734 (1991).
11. Sillen, L. G., and A. E. Martell. *Stability Constants of Metal-Ion Complexes,* Special Publication No. 17 (London: The Chemical Society, Burlington House, 1964).
12. Ehrlich, H. L. *Geomicrobiology,* 2nd ed. (New York: Marcel Dekker, 1990).
13. Lazaroff, N., W. Sigal, and A. Wasserman. "Iron Oxidation and Precipitation of Ferric Hydroxysulfates by Resting *Thiobacillus ferrooxidans* Cells," *Appl. Environ. Microbiol.* 43:924–938 (1982).
14. Lazaroff, N., L. Melanson, E. Lewis, N. Santoro, and C. Pueschel. "Scanning Electron Microscopy and Infrared Spectroscopy of Iron Sediments Formed by *Thiobacillus ferrooxidans,*" *Geomicrobiol. J.* 4:231–268 (1985).
15. Ghiorse, W. C. "Biology of Iron- and Manganese-Depositing Bacteria," *Annu. Rev. Microbiol.* 38:515–550 (1984).
16. Luetters, S., and H. H. Hanert. "The Ultrastructure of Chemolithoautotrophic *Gallionella ferruginea* and *Thiobacillus ferrooxidans* as Revealed by Chemical Fixation and Freeze-Etching," *Arch. Microbiol.* 151:245–251 (1989).

17. Nielsen, A. M., and J. V. Beck. "Chalcocite Oxidation and Coupled Carbon Dioxide Fixation by *Thiobacillus ferrooxidans*," *Science* 175:1124–1126 (1972).

18. Duncan, D. W., J. Landesman, and C. C. Walden. "Role of *Thiobacillus ferrooxidans* in the Oxidation of Sulfide Minerals," *Can. J. Microbiol.* 13:397–403 (1967).

19. Oremland, R. S., J. T. Hollibaugh, A. S. Maest, T. S. Presser, L. G. Miller, and C. W. Culbertson. "Selenate Reduction to Elemental Selenium by Anaerobic Bacteria in Sediments and Culture: Biogeochemical Significance of a Novel, Sulfate-Independent Respiration," *Appl. Environ. Microbiol.* 55:2333–2343 (1989).

20. Lovley, D. R., E. J. P. Phillips, Y. A. Gorby, and E. R. Landa. "Microbial Reduction of Uranium," *Nature (London)* 350:413–416 (1991).

21. Bopp, L. H., and H. L. Ehrlich. "Chromate Resistance and Reduction in *Pseudomonas fluorescens* Strain LB300," *Arch. Microbiol.* 150:426–431 (1988).

22. Romanenko, V. I., and V. N. Korenkov. "A Pure Culture of Bacteria Utilizing Chromates and Biochromates as Hydrogen Acceptors in Growth Under Anaerobic Conditions," *Mikrobiologiya* 46:414–417 (1977).

23. Wang, P. C., T. Mori, K. Komori, M. Sasatsu, K. Toda, and H. Ohtake. "Isolation and Characterization of an *Enterobacter cloacae* Strain that Reduces Hexavalent Chromium Under Anaerobic Conditions," *Appl. Environ. Microbiol.* 55:1665–1669 (1989).

24. Fedorak, P. M., D. W. S. Westlake, C. Anders, B. Kratochvil, N. Motkosky, W. B. Anderson, and P. M. Huck. "Microbial Release of $^{226}Ra^{2+}$ from $(Ba,Ra)SO_4$ Sludges from Uranium Mine Wastes," *Appl. Environ. Microbiol.* 52:262–268 (1986).

25. Stevenson, F. J., and A. Fitch. "Chemistry of Complexation of Metal Ions with Soil Solution Organics," in *Interactions of Soil Minerals with Natural Organics and Microbes. SSSA Special Publication Number 17*, P. M. Huang and M. Schnitzer, Eds. (Madison, WI: Soil Science Society of America, Inc., 1986), pp. 29–58.

26. Wildung, R. E., T. R. Garland, and H. Drucker. "Nickel Complexes with Soil Microbial Metabolites — Mobility and Speciation in Soils, " in *Chemical Modeling in Aqueous Systems. American Chemical Society Symposium Series No. 93* (Washington, DC: American Chemical Society, 1979), pp. 181–200.

27. Samanidou, V., and K. Fytianos. "Mobilization of Heavy Metals from River Sediments of Northern Greece by Complexing Agents," *Water, Air, and Soil Pollut.* 52:217–225 (1990).

28. Francis, A. J., C. J. Dodge, and J. B. Gillow. "Biodegradation of Metal Citrate Complexes and Implications for Toxic-Metal Mobility," *Nature* 356:140–142 (1992).

29. Francis, A. J., and C. J. Dodge. "Influence of Complex Structure on the Biodegradation of Iron-Citrate Complexes," *Appl. Environ. Microbiol.* 59:109–113 (1993).

30. Huber, A. L., B. E. Holbein, and D. K. Kidby. "Metal Uptake by Synthetic and Biosynthetic Chemicals," in *Biosorption of Heavy Metals*, B. Volesky, Ed. (Boca Raton, FL: CRC Press, 1990), pp. 249–292.

31. Hider, R. C. "Siderophore Mediated Absorption of Iron," *Struct. Bonding (Berlin)* 58:25–87 (1984).

32. Robinson, J. B., and O. H. Tuovinen. "Mechanism of Microbial Resistance and Detoxification of Mercury and Organo-Mercury Compounds: Physiological, Biochemical, and Genetic Analyses," *Microbiol. Rev.* 48:95–124 (1984).

33. Brierley, C. L. "Metal Immobilization Using Bacteria," in *Microbial Mineral Recovery*, H. L. Ehrlich and C. L. Brierley, Eds. (New York: McGraw-Hill, 1990), pp. 303–323.

34. Macaskie, L. E., and A. C. R. Dean. "Cadmium Accumulation by a *Citrobacter* sp.," *J. Gen. Microbiol.* 130:53–62 (1984).

35. Aiken, R. M., and A. C. R. Dean. "Lead Accumulation by *Pseudomonas fluorescens* and by a *Citrobacter* sp.," *Microbios Lett.* 9:55–66 (1978).

36. Macaskie, L. E., and A. C. R. Dean. "Use of Immobilized Biofilm of *Citrobacter* sp. for the Removal of Uranium and Lead from Aqueous Flows," *Enzyme Microbiol. Technol.* 9:2–4 (1987).

37. Macaskie, L. E., and A. C. R. Dean. "Uranium Accumulation by Immobilized Biofilms of a *Citrobacter* sp.," in *Biohydrometallurgy*, P. R. Norris and D. P. Kelly, Eds. (Kew, Surrey, U.K.: Science and Technology Letters, 1988), pp. 556–557.

38. Lucas, J., and L. Prevot. "Synthése de L´apatite par Voie Bacterienne a Partir de Matière Organique Phosphatée et de Divers Carbonates de Calcium dans des Eaux Douce et Marine Naturelles," *Chem. Geol.* 42:101–118 (1984).

39. Hirschler, A., J. Lucas, and J.-C. Hubert. "Bacterial Involvement in Apatite Genesis," *FEMS Microbiol. Ecol.* 73:211–220 (1990).

40. Hirschler, A., J. Lucas, and J.-C. Hubert. "Apatite Genesis: A Biologically Induced or Biologically Controlled Mineral Formation Process?" *Geomicrobiol. J.* 7:47–57 (1990).

41. Pettersson, A., L. Kunst, B. Bergman, and G. M. Roomans. "Accumulation of Aluminum by *Anaboena cylindrica* into Polyphosphate Granules and Cell Walls: An X-ray Energy Dispersive Microanalysis Study," *J. Gen. Microbiol.* 131:2545–2548 (1985).

42. Volesky, B., "Removal and Recovery of Heavy Metals by Biosorption," in *Biosorption of Heavy Metals*, B. Volesky, Ed. (Boca Raton, FL: CRC Press, 1990), pp. 7–43.

43. Beveridge, T. J., and R. J. Doyle, Eds. *Metal Ions and Bacteria* (New York: Wiley, 1989).

44. Emerson, W. W., R. C. Foster, and J. M. Oades. "Organo-Mineral Complexes in Relation to Soil Aggregation and Structure," in *Interactions of Soil Minerals with Natural Organics and Microbes. SSSA Special Publication Number 17*, P. M. Huang and M. Schnitzer, Eds. (Madison, WI: Soil Science Society of America, Inc., 1986), pp. 521–548.

45. Ehrlich, H. L., and C. L. Brierley, Eds. *Microbial Mineral Recovery* (New York: McGraw-Hill, 1990).

46. Wong, P. T. S., Y. K. Chau, and P. L. Luxon. "Methylation of Lead in the Environment," *Nature (London)* 253:263–264 (1975).

47. Thayer, J. S., and F. E. Brinckman. "The Biological Methylation of Metals and Metalloids," *Adv. Organometallic Chem.* 20:313–356 (1982).

48. Thayer, J. S., G. J. Olson, and F. E. Brinckman. "Iodomethane as a Potential Metal Mobilizing Agent in Nature," *Environ. Sci. Technol.* 18:726–729 (1984).

49. Brinckman, F. E., and G. J. Olson. "Chemical Principles Underlying Bioleaching of Metals from Ores and Solid Wastes, and Bioaccumulation of Metals from Solution," in *Workshop on Biotechnology for the Mining, Metal-Refining and Fossil Fuel Processing Industries. Biotech. Bioeng. Symp. No. 16*, H. L. Ehrlich and D. S. Holmes, Eds. (New York: Wiley, 1986), pp. 35–44.

50. White, R. H. "Analysis of Dimethyl Sulfonium Compounds in Marine Algae," *J. Marine Res.* 40:529–536 (1982).

51. Manley, S. L., and M. N. Dastoor. "Methyl Iodide (CH_3I) Production by Kelp and Associated Microbes," *Mar. Biol. (Berlin)* 98:477–482 (1988).

52. Harper, D. B. "Halomethane from Halide Ion — A Highly Efficient Fungal Conversion of Environmental Significance," *Nature (London)* 315:55 (1985).

53. Cork, D. J. "Microbial Conversion of Sulfate to Sulfur — An Alternative to Gypsum Synthesis," *Adv. Biotechnol. Process.* 4:183–209 (1985).

54. Ishikawa, T., T. Murayama, I. Kawahara, and T. Imaizumi. "A Treatment of Acid Mine Drainage Utilizing Bacterial Oxidation," in *Recent Progress in Biohydrometallurgy*, G. Rossi and A. E. Torma, Eds. (Iglesias, Italy: Associazone Mineraria Sarda, 1983), pp. 393–407.

55. Imaizumi, T. "Some Industrial Applications of Inorganic Microbial Oxidation in Japan," in *Workshop on Biotechnology for the Mining, Metal-Refining and Fossil Fuel Processing Industries. Biotech. Bioeng. Symp. No. 16*, H. L. Ehrlich and D. S. Holmes, Eds. (New York: Wiley, 1986), pp. 363–371.

56. Whitlock, J. L. "Biological Detoxification of Precious Metal Processing Wastewaters," *Geomicrobiol. J.* 8:241–249 (1990).

57. Whitlock, J. L., and T. I. Mudder. "The Homestake Wastewater Treatment Process: Biological Removal of Toxic Parameters from Cyanidation Wastewaters and Bioassay Effluent Evaluation," in *Fundamental and Applied Biohydrometallurgy*, R. W. Lawrence, R. M. R. Branion, and H. G. Ebner, Eds. (Amsterdam: Elsevier, 1986), pp. 327–339.

58. Whitlock, J. L., and G. R. Smith. "Operation of Homestake's Cyanide Biodegradation Wastewater System Based on Multi-Variable Trend Analysis," in *Biohydrometallurgy: Proceedings of the International Symposium, Jackson Hole, Wyoming, August 13–18, 1989, CANMET SP89–10*, J. Salley, R. G. L. McCready, and P. L. Wichlacz, Eds. (Ottawa: Canada Centre for Mineral and Energy Technology, 1989), pp. 613–625.

59. Darnall, D. W., R. M. McPherson, and J. Gardea-Torresdey. "Metal Recovery from Geothermal Waters and Groundwaters Using Immobilized Algae," in *Biohydrometallurgy: Proceedings of the International Symposium, Jackson Hole, Wyoming, August 13–18, 1989, CANMET SP89–10*, J. Salley, R. G. L. McCready, and P. L. Wichlacz, Eds. (Ottawa: Canada Centre for Mineral and Energy Technology, 1989), pp. 341–348.

60. Greene, B., and D. W. Darnall. "Microbial Oxygenic Photoautotrophs (Cyanobacteria and Algae) for Metal-Ion Binding," in *Microbial Mineral Recovery*, H. L. Ehrlich and C. L. Brierley, Eds. (New York: McGraw-Hill, 1990), pp. 277–302.

61. Brierley, J. A., C. L. Brierley, and G. M. Goyak. "A New Wastewater Treatment and Metal Recovery Technology," in *Fundamental and Applied Biohydrometallurgy*, R. W. Lawrence, R. M. R. Branion, and H. G. Ebner, Eds. (Amsterdam: Elsevier, 1985), pp. 291–303.

62. Gale, N. L., and B. G. Wixson. "Removal of Heavy Metals from Industrial Effluents by Algae," *Dev. Ind. Microbiol.* 20:259–273 (1979).

63. Gale, N. L. "The Role of Algae and Other Microorganisms in Metal Detoxification and Environmental Clean-Up," in *Workshop on Biotechnology for the Mining, Metal-Refining and Fossil Fuel Processing Industries. Biotech. Bioeng. Symp. No. 16*, H. L. Ehrlich and D. S. Holmes, Eds. (New York: Wiley, 1986), pp. 171–180.

64. Kleinmann, R. L. "Biological Treatment of Mine and Mineral Processing Wastewater," in *Biohydrometallurgy: Proceedings of the International Symposium, Jackson Hole, Wyoming, August 13–18, 1989, CANMET SP89–10*, J. Salley, R. G. L. McCready, and P. L. Wichlacz, Eds. (Ottawa: Canada Centre for Mineral and Energy Technology, 1989), pp. 593–598.

65. Hellier, W. W., Jr. "Constructed Wetlands in Pennsylvania an Overview," in *Biohydrometallurgy: Proceedings of the International Symposium, Jackson Hole, Wyoming, August 13–18, 1989, CANMET SP89–10*, J. Salley, R. G. L. McCready, and P. L. Wichlacz, Eds. (Ottawa: Canada Centre for Mineral and Energy Technology, 1989), pp. 599–611.

mental factors that affect degradation are important criteria for determining the treatment approach. The behavior of oil spilled in water is considerably different than the behavior of oil spilled in terrestrial environments.[26] Because oil is immiscible with water, it tends to spread in a horizontal fashion in aquatic systems. The characteristic high surface to volume ratio in slicks renders the oil susceptible to volatilization, dissolution, emulsification by wind and wave action, photooxidation, and biodegradation. Biodegradation of oil competes with evaporative losses and photooxidation.

Chlorinated solvents are degraded by biological and abiological transformations.[27] Abiotic transformations of chlorinated compounds are relatively slow. However, given the lengthy times associated with subsurface transport, the impact of abiotic transformation on the degradation of chlorinated solvents can be significant.[27] Transformations mediated by microorganisms can be relatively rapid if sufficient organic substrates, nutrients, and electron acceptors are present to support microbial activity. Complete mineralization of chlorinated aliphatics in the environment most likely occurs by the combined activities of anaerobic and aerobic microorganisms and abiotic transformations.

2.2 Bioreactor Designs

Bioreactors, which include treatment lagoons and ponds, composting, and landfarming can be used for the biological treatment of hazardous organic materials. These methods provide varying degrees of control over the parameters necessary to promote microbial activity.[28,29] And while increased process control can potentially decrease treatment times, it does result in increased capital and operating costs. Each treatment approach must also consider the physical limitations of contacting the contaminant in the form of liquids, solids, and gases with the microorganism.

Bioreactors may use freely suspended biomass as in slurry or tank reactors or use immobilized biomass as fixed film bioreactors.[30,31] Tank bioreactors, operated in batch or continuous modes, have been used singly or in series, aerobically or anaerobically.[28,29] Examples (Figures 1 and 2) include continuous stirred tank reactor (CSTR), fed batch reactor (FBR), and sequencing batch reactor (SBR). While the designs of these reactors are similar, they differ in the manner the waste is introduced. For the CSTR, the liquid stream containing the waste is continually introduced, while a stream with cell biomass continually leaves. Biomass should be in a single physiological state throughout the operation. The CSTR is excellent for inhibitory wastes. Biomass is only exposed to low contaminant concentrations.

There is no flow in a FBR. In a FBR, a constant stream of contaminants can enter, but there is no outward flow. Unless FBRs are controlled properly, the physiological condition of the cells as well as the reaction rates will change with time. The SBR requires several in parallel: one is being filled; one is in operation, and yet another is settling biomass and the effluent pumped out. The biomass in the SBR at time zero is exposed to high, potentially inhibiting concentrations of

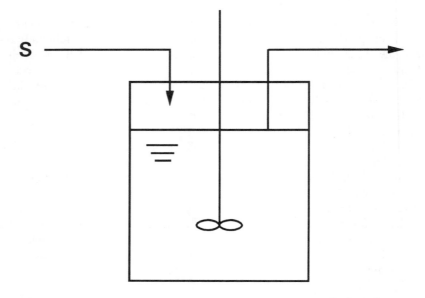

Figure 1. Schematic diagram of a continuous stirred tank reactor operated singly (A, B), in series (C, D), without (A,C) and with (B,D) recirculation. Recirculation can occur before or after cell separation (CS). Redrawn from Grady and Lim.[28]

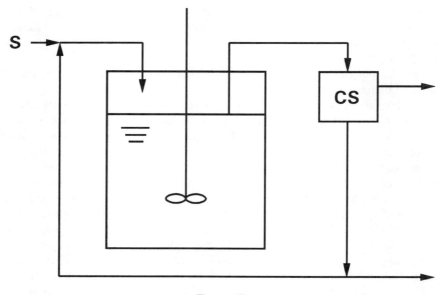

Figure 1B.

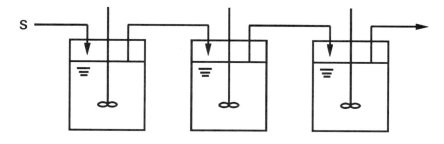

Figure 1C.

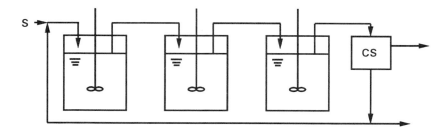

Figure 1D.

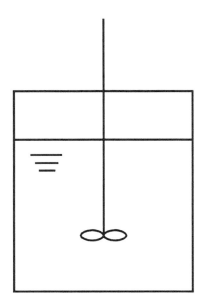

Figure 2.   Schematic of a batch (A) and a fed batch (B) reactor. Redrawn from Grady and Lim.[28]

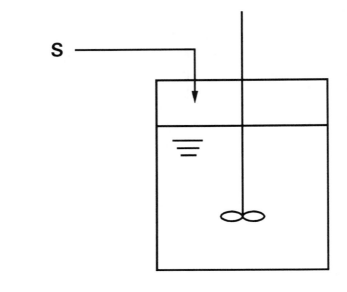

Figure 2B.

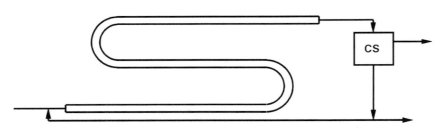

Figure 3. Schematic of a plug flow reactor. Redrawn from Grady and Lim.[28]

organics. In contrast, the FBR is excellent for extremely concentrated, even nonaqueous feeds. Wastes can be added at a rate to maintain concentrations below inhibitory levels. A FBR is not appropriate for low concentrations. If fed with a dilute waste, the FBR will overflow too quickly.

A plug flow reactor (Figure 3) can be configured as a simple tube, or it can contain a packing.[28] A plug flow reactor is comparable to a batch reactor, even though there is a continual flow of material into and out of the reactor. The physiological condition of the cells as well as the reaction rates will be different at the different points along the flow path. A steady state, however, can be achieved at any point along the reactor. A plug flow reactor is not good for inhibitory organics as cells near the influent are exposed to high feed concentrations.

Fixed film bioreactors can be configured as stationary particle reactors such as hollow fiber reactors and packed bed systems (Figure 4), moving surface reactors

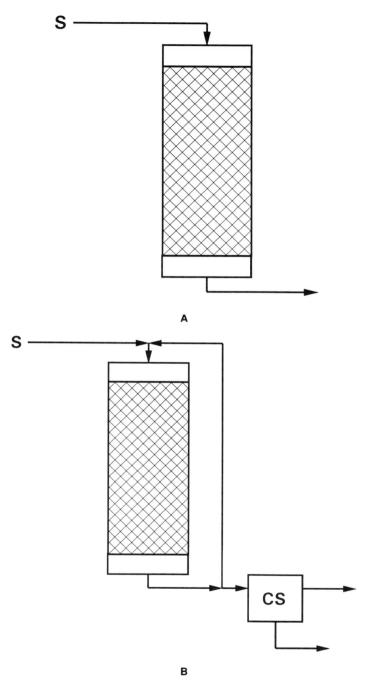

Figure 4.   Schematic of trick bed bioreactors without (A) and with (B) recirculation. Recirculation can occur before or after CS. Redrawn from Grady and Lim.[28]

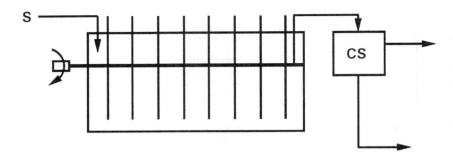

Figure 5. Schematic of a rotating biological contactor. Redrawn from Grady and Lim.[28]

such as rotating biological contactor (RBC) (Figure 5), and mixed particle reactors such as stirred tank and fluidized bed reactors.[28,31] For fixed film bioreactors, the biomass adheres to a solid support. In packed tower reactors, which are variations on the classical trickling filter, liquid forms a thin film as it trickles down the immobilized support material.[28] In this reactor design, there is a considerable change in reactor conditions from one end of the column to the other. This difference can be minimized by recycling the fluid. A fixed film reactor with a high recycle ratio is functionally identical to a CSTR. For RBC, the microorganisms adhere to a series disk mounted on a common axis.[28,29] Disks are continually rotated within the tank such that half of the disk is submerged at any one time. Rotating biological contactors can be linked in parallel or in series.

Fluidized beds are column reactors in which the wastewater is pumped upward through a particulate bed of material upon which a biofilm has developed.[29] Fluidized beds can be aerated directly or in the recycle loop. The important advantage of the fluidized bed is the very high surface area of biofilm per unit volume of reactor that can be achieved by the use of small particles. Trickling filters with advanced plastic packings cannot achieve these ratios, and pack beds constructed with small particles will quickly plug.

Composting and landfarming traditionally used for the decomposition of bulk organic materials are being applied to hazardous waste management and environmental remediation.[32,33] The composting process begins with the mixing of sewage sludge, bulking agents, nutrients, water, and the hazardous material. The pile is aerated and can be manipulated to control the heat generated by metabolic activity. The active stage of composting can take place by in vessel, static pile, turned windrow, or sheet processes. In vessel procedures are often preferred for the treatment of hazardous waste because of environmental regulations, increased process control, and improved control of off gases.[31] Landfarming utilizes indigenous soil microorganisms to degrade or transform hazardous organic materials.[7] The landfarm site is managed to promote microbial activity by tilling, pH control, and fertilization. Tilling aerates the soil and distributes the organics evenly. The site is also controlled to prevent overloading with organics and inhibition of microbial activity.

3. EXAMPLES OF BIOLOGICAL TREATMENT PROCESSES

3.1 Treatment of Liquid Wastes

3.1.1 In Situ Treatment for the Remediation of Surface and Groundwater

Oil spills in oceans are treated with a variety of mechanical, chemical, and biological methods depending on the type of oil, location (i.e., proximity to land, sensitive marine areas), and sea and weather conditions.[8] Treatment of large oil spills as exemplified by the *Exxon Valdez* spill is initiated by physical containment and free product recovery.[9] Shoreline remediation can employ mechanical methods such as water washing or steam cleaning to lift and move the oil. Dispersants such as Corexit 9580 M2 can be used to disrupt the oil slick in order to enhance dissolution and biodegradation. Bioremediation of large surface oil spills is not considered an established technology and is usually viewed as a polishing step or for long-term remediation of oil-contaminated shorelines.[8,9]

The microbial degradation of oil in marine environments, which is limited by the availability of nitrogen and phosphorous, can be stimulated by the application of fertilizers.[34,35] Fertilizers such as Inipol, a slow release oleophilic fertilizer, IBDU briquets composed of isobutyldiene diurea, and a granulated fertilizer have been tested on the oil-contaminated shorelines of Prince William Sound and shown to enhance biodegradation.[8,36] Due to the required biological treatment times, mechanical treatment followed by bioremediation was recommended for heavily contaminated areas.

Mechanical oil recovery, followed by venting and *in situ* bioremediation was used to remediate soils contaminated with diesel oil.[15] Skimmers in groundwater draw down wells in Australia were used to retrieve 135,000 of the 185,000 liters of diesel oil spilled in an area of 2000 square meters. The initial skimming operation removed approximately half of the oil load in the soil, thus, decreasing contaminant concentrations from 20,000 mg/kg to 10,000 mg/kg. In a pilot scale operation, venting and venting with nutrient amendment resulted in further removal of diesel oil from the subsurface: 13% decrease in 6 months followed by an 18% decrease in a subsequent 6 month period. Aeration was provided by a vacuum system installed in boreholes, while nitrogen and phosphorous were added via a drip irrigation system installed just below the surface of the soil.

3.1.2 Bioreactor Treatment of Contaminated Water

While *in situ* remediation of contaminated aquifers represents the least cost biological treatment alternative, above ground bioreactors may be preferred in some cases as they offer better process control together with more effective containment of toxic degradation products. Groundwater may be pumped to the surface for treatment in bioreactors.

A combination of *in situ* and above ground biological treatment with indigenous microorganisms was used to remediate a site contaminated with ethylene glycol.[4] A preliminary assessment at one site indicated the presence of a microbial population capable of degrading ethylene glycol. This microbial activity in the subsurface soil/groundwater environments and the above ground bioreactor was enhanced by the addition of oxygen, lime to adjust the pH, and diammonium phosphate as the nitrogen and phosphorous source. Recovery wells were used to withdraw contaminated groundwater for treatment in an above ground bioreactor. Following treatment, the water was reinjected into the ground in a three-phase system. Injection systems were used to: (1) flush contaminated soil and transmit the water to the recovery wells; (2) enhance microbial growth through nutrient, lime, and oxygen amendment; and (3) through surface application, flush the vadose zone and enhance microbial growth in this region. Thus, the whole system above and below ground formed a looped system. At the start of treatment, the initial groundwater concentration of ethylene glycol was 1440 ppm, and within 26 days of treatment, concentrations decreased to <50 ppm, which was the detection limit of the analytical method utilized. The remaining pockets of contamination, primarily in the lagoon spill area and contaminant plume, were treated with surface amendment with lime and diammonium phosphate.

In a remediation program similar to that described above for the ethylene glycol-contaminated site, a modified activated sludge system was used to treat groundwater contaminated with methylene chloride.[4,5] The bulk of the free material spilled when a pipeline ruptured was recovered from the soil by using trenches and a vacuum system. The majority of the groundwater contamination was removed by air-stripping, and the residual contaminants were treated biologically. For the first full scale field operation, bacteria from the plant wastewater system were used as the inoculum to start the above ground biological treatment system and introduced into the subsurface environment by the injection wells. Feasibility studies indicated that these microorganisms, which had been continually exposed to low levels of methylene chloride, were capable of oxidizing methylene chloride to chloride and carbon dioxide. Soil microorganisms, which had been exposed to methylene chloride for a relatively short period of time, did not metabolize this contaminant. However, by the second full scale field operation initiated 4 years after the discovery of the spill, soil microorganisms had acquired the ability to degrade methylene chloride.

A fixed film aerobic biodegradation system was tested for the remediation of groundwater with low levels of benzene, toluene, xylene, and petroleum hydrocarbons.[37] The column was packed with a structured polyvinyl chloride (PVC) medium, and the column operated in the downflow mode, while air flowed countercurrently upward from the bottom. The structured packing provided a stable matrix for biofilm development and allowed air passage necessary to maintain aerobic conditions. The groundwater was supplemented, throughout the year long operation, with a nutrient feed.

When the hazardous material being treated does not serve as the primary growth substrate, the treatment process design must provide for adequate carbon for growth and energy requirements of the microorganism. For example, trichloroethylene does not serve as a primary growth substrate. However, this compound can be biodegraded under aerobic conditions through the cometabolic processes.[38–41] Methanotrophic bacteria can cometabolize chlorinated solvents using methane to satisfy their carbon and energy requirements through the oxidation of methane. The enzyme system methane monooxygenase is responsible for both methane and trichloroethylene (TCE) oxidation.[39] Although the prime catabolic function of methane monooxygenase (MMO) is to catalyze the conversion of methane to methanol, its low substrate specificity enables it to mediate TCE oxidation.

Although methane is needed for enzyme induction, Speitel and Leonard[42] have utilized methanotroph physiology in developing a sequencing biofilm reactor that minimizes competitive inhibition from methane[43,44] and increases chlorinated solvent degradation rates. The bioreactor design employs two modes of operation consisting of a growth mode and a degradation mode. During the growth mode (Figure 6A), methane and oxygen were supplied in the gas phase to the organisms. This allowed for increased biomass in the reactor and induction of nonspecific enzymes essential for solvent degradation. The degradation mode (Figure 6B) involved filling the bioreactor with water containing formate as a source of energy for reducing power and chloroform as a model chlorinated compound. Although this reactor design and operation have not been optimized, the authors demonstrated significantly better results with the sequencing bioreactor than a packed-bed continuous flow reactor due primarily to the increased biomass permitted in the sequencing reactor.

Gas-tight rotating biological contactors were evaluated for the cometabolic treatment of chlorinated solvents using methanotrophic bacteria[45] and for the treatment of 2,4 dichlorophenol using the fungus *Phanerochaete chrysosporium*.[46] In this design, the microorganisms colonize polyethylene disks housed in a gas-tight chamber. Methane and glucose were the cosubstrates for the methanotrophs and the fungus, respectively. The gas-tight feature of this type of bioreactor is beneficial for a number of reasons. This design minimizes the air-stripping of volatile organic compounds (VOC) and, in the case of the methanotrophic bacteria, loss of the cosubstrate methane. Thus, the enclosed design maintains the contact of the VOC and volatile cosubstrates with the microorganisms. The bioreactor used for evaluating the methanotrophic bacteria was effective in treating VOC concentrations of 1 to 500 ppm.[45] In this case, a viable biofilm was maintained at an organic concentration of 1 ppm or less. For the fungi, the bioreactor was operated in batch mode and evaluated at 3 concentrations of dichlorophenol — 20, 50, and 100 mg/mL.[46] A final concentration of <1 mg/mL was achieved by 24 h for the 20 and 50 mg/mL starting concentration and at 120 h for the 100 mg/mL starting concentration.

An aerobic pilot scale trickling filter packed-bed reactor was field tested for the remediation of TCE-contaminated groundwater.[6] Two reactor units were connected in series and inoculated with a methanotrophic consortia, which cometabolized TCE using methane as a cosubstrate. The reactor was supplemented with nitrate, phosphate, sulfate, and trace minerals. TCE contaminated groundwater (0.5 to 1.4 mg TCE per liter) was added at a rate of 2 L/min to the top of the bioreactor, and methane was introduced at the top and midpoint of the reactor units. The percentage of TCE loss due to air-stripping decreased with increasing TCE concentration and increasing biodegradation rates. The bioreactor system recovered quickly from operational upsets and system shutdowns that lasted for as long as 5 days.

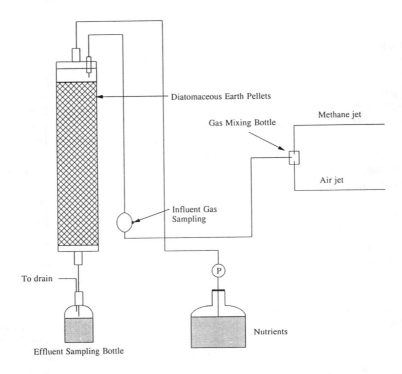

Figure 6. Schematic of sequencing bioreactor configured for (A) growth mode and (B) degradation mode. Redrawn from Speitel and Leonard.[42]

The use of sequential anaerobic and aerobic environments for the treatment of wastewater containing highly chlorinated or complex mixtures of contaminants is proving to be an effective treatment approach.[47–49] The sequential treatment of wastewater under anaerobic and aerobic conditions can be more successful than a wholly aerobic or anaerobic process, e.g., the biological degradation of mixtures of highly chlorinated hydrocarbons.[48] While anaerobic reductive dechlorination can result in partial dechlorination, reaction rates decrease with decreased chlorinated substituents. Thus, partially dechlorinated metabolites tend to accumulate. Aerobic microorganisms, in contrast, can readily degrade these less substituted chlorinated compounds, but are ineffective against the highly substituted compounds. Therefore, the sequential treatment of highly chlorinated hydrocarbons under anaerobic conditions followed by aerobic conditions results in the complete destruction of these compounds.

Fathepure and Vogel[48] evaluated two sequencing, anaerobic-aerobic continuous fed, upflow, biofilm reactors for the treatment of chlorinated hydrocarbons. Biofilms were established in the anaerobic and aerobic bioreactors by seeding columns with microorganisms that originated from a primary anaerobic digestor

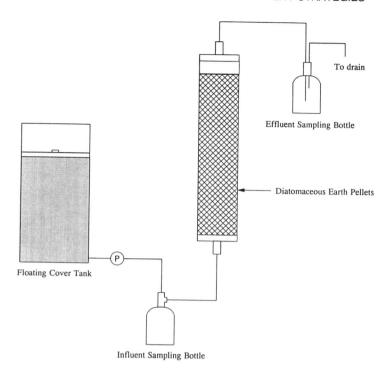

Figure 6B.

and an activated sludge system, respectively. Columns were acclimated and tested separately, then connected and evaluated in series so that the effluent from the anaerobic bioreactor was pumped directly into the aerobic bioreactor. The anaerobic bioreactor was acclimated to the parent compounds hexachlorobenzene, tetrachloroethylene, and chloroform. The aerobic bioreactor was acclimated to the partially dechlorinated metabolites expected to emerge from the anaerobic treatment, i.e., dichlorobenzenes, dichloroethylenes, and methylene chloride. Results demonstrated that for sequential bioreactor operation overall conversion rates to CO_2 were 61%, 49%, and 22% for chloroform, trichloroethylene, and hexachlorobenzene, respectively. Only 7%, 1%, and 16% of this conversion had occurred in the anaerobic reactor. Conversion to nonvolatile compounds was 22%, 47%, and 71% for chloroform, trichloroethylene, and hexachlorobenzene, respectively.

Mueller et al.[49] evaluated a 454 L capacity sequential bioreactor system for the treatment of groundwater contaminated with creosote and pentachlorophenol. The treatment system consisted of two aerobic reactors operated in series. The first reactor, which was continuously fed with nutrient-amended contaminated groundwater, was acclimated with a combination of 13 microbial strains, which degraded the majority of creosote constituents. The second reactor, operated in batch mode, was fed the effluent of the first reactor and inoculated daily with two microorganisms *Pseudomonas paucimobilis* strain EPA 505 and *Pseudomonas* sp. strain SR3. Strain EPA 505 degraded high molecular weight polycyclic aromatic hydrocarbons, and strain SR3 degraded pentachlorophenol. During operation at a feed rate of 114 L/day, creosote constituents decreased from 1000 mg/L to less than 9 mg/L.

3.2 Treatment of Solids

The biological treatment of solid phase hazardous waste can be accomplished by landfarming or composting, by enclosed bioreactors, or by utilizing physical treatment in combination with biological treatment. Again, *in situ* bioremediation offers a low cost remediation option. Yet, effective treatment by *in situ* bioremediation may be limited by the ability to promote bacterial activity. To overcome these limitations, the use of bioreactors as simple as landfarming or composting may be effective for the treatment of soils. Enclosed bioreactors may be used in situations requiring increased emission control or to maintain optimal operating conditions for enhanced microbial activity or decreased treatment times.[50]

3.2.1 In Situ Treatment for the Remediation of Contaminated Soils

In situ bioremediation of contaminated subsurface materials may be used as a stand alone process in combination with a physical treatment such as the mechanical retrieval or removal of free contaminants. The microbial degradation of petroleum in subsurface environments is often limited by the availability of nitrogen, phosphorus, and oxygen. Oxygen is required for aerobic respiration and for the enzymatic oxidation of hydrocarbons.[51,52] Biodegradation can occur with nitrate or sulfate reduction respiration with alternative mechanisms of degradation.[53]

Because the availability of oxygen is often the limiting factor for *in situ* bioremediation of petroleum-contaminated subsurface soils, humidified air can be used to stimulate *in situ* biodegradation of gasoline and diesel.[54] One site, contaminated to a depth of 60 ft and a radius of 60 ft, had initial gasoline concentrations of 3000 to 4000 ppm and diesel concentrations ranging from 10,000 to 14,000 ppm. Following treatment, gasoline concentrations had decreased to close to detection limits, and diesel concentrations were decreased by half. An increase in carbon dioxide in off-gas collected at wells located at the perimeter of the site was attributed to microbial degradation. Nutrient amendment was not necessary. Because of the sites proximity to agricultural activities, sufficient levels of nitrogen and phosphorous were present in the soils to support microbial activity.

Hydrogen peroxide may also be used to introduce oxygen into subsurface environments. Hydrogen peroxide was evaluated as the oxygen source for the bioremediation of petroleum-contaminated sites.[55-57] In one case, injection wells were used to raise the water table to the level of the contamination by pumping in clean water. The hydrogen peroxide and ammonium and phosphate salts pumped into the injection wells increased the oxygen and nutrient levels in this, now, saturated zone.[55,56] The breakthrough of oxygen in monitoring wells corresponded to the depletion of BTEX (benzene, toluene, ethyl benzene, and xylenes).[56]

Following physical cleaning methods, *in situ* biodegradation enhanced by the addition of nutrients and hydrogen peroxide was used to treat residual material at a site contaminated with gasoline.[57] Initial treatment involved skimming and air-stripping to recover free products from the groundwater, and soil venting by vacuum extraction was used to treat the vadose zone. Hydrogen peroxide was added to the subsurface environment a month after nutrient amendment had been initiated. It was estimated that 72% of the 24,300 kg of gasoline was removed by bioremediation and that the combination of physical and biological treatment techniques was successful in decreasing the amounts of gasoline in the aquifer to 10 mg/kg.

3.2.2 Soil Composting and Land Treatment

Traditional approaches to *in situ* biological treatment (e.g., soil nutrient and/or biomass additions) may generate toxic chemical metabolites. For example, the accumulation of vinyl chloride can result from partial dechlorination of chlorinated compounds such as perchloroethylene (PCE) and TCE.[58] Similarly, pesticides may also undergo incomplete conversion in the environment,[59,60] and acetone was postulated to be a metabolite resulting from the degradation of isopropanol at a contaminated site.[5]

Increased process control may be achieved by composting and landfarming of contaminated soils, both of which are accepted waste treatment technologies.[7,32,33,61] As for *in situ* remediation of oil-contaminated environments, the increased carbon load in landfarming due to the contaminant must be balanced by the application of nitrogen and phosphorous.[7,62] Sufficient aeration can usually be achieved by tilling, and the pH adjusted by the addition of lime. Soil temperature, which impacts the biodegradation rates, will influence the rate at which the contaminants can be applied. Thus, reapplication rates need to be seasonally adjusted; more frequent applications are needed during the summer than in the winter. Temperature is also an important parameter in composting as it impacts the biodegradation and volatilization of the organic.[61] Composting is characterized by self-heating due to metabolic activity. Yet, the temperature can be controlled somewhat by manipulating the piles.

Landfarming, traditionally used as a disposal technique, has been used successfully to remediate soils contaminated with petroleum[7,17,62,63] and chlorinated hydrocarbons.[64] Following excavation of heavily contaminated soil, a program of tilling, limestone, and fertilizer application was used to promote microbial degradation of kerosene spilled in an agricultural area. Tilling was used to aerate the soil as well as distribute the lime and fertilizer. After 1 year, germination rate and yields in the affected field were comparable to uncontaminated areas, and by 2 years, negligible amounts of kerosene remained.

Aerobic composting[33] using the white rot fungus *Phanerochaete chrysosporium* may be used to treat soils contaminated with hazardous organics.[65,66] This fungus cometabolizes organic pollutants with the nonspecific ligninase enzymes.[65,66] *P. chrysosporium* and other white rot fungi are able to degrade a range of organic compounds from the petrochemical industry.[67,68] *P. chrysosporium* is known to cometabolize a variety of pesticides such as chlordane, lindane, 2,4,5 T, and DDT.[65,66,69]

White rot fungi *P. chrysosporium* and *P. sordida* were field tested for the ability to deplete pentachlorophenol (PCP) in soils contaminated with wood preservative.[64] The study area was tilled to equalize the concentrations of PCP and to promote the volatilization of mineral spirits previously shown to inhibit fungal growth. Soils were sterilized by fumigation with a mixture of methyl bromide and chloropicrin. Some plots to be seeded with the fungi were amended with wood chips and peat to permit the growth of these microorganisms. The PCP concentrations in soils seeded with fungi and amended with peat and chips decreased approximately 90% over a 6.5 week period. In comparison, PCP levels decreased approximately 20% in soils amended with chips and/or peat or soils receiving no treatment. While striking decreases in PCP concentrations were observed for fungi seeded soils, the authors were cautious in interpreting the fate of this compound. Most of the PCP was metabolized to nonextractable soil-bound products. Only a small percentage of PCP loss was due to mineralization or volatilization.

3.2.3 Bioreactors for the Treatment of Solids

As described above and in other cases, composting and landfarming can result in the accumulation of recalcitrant or partially degraded materials.[7,63,64,66,70] The principal advantage of enclosed reactor systems over *in situ*, land treatment and composting systems is better process control, which can result in controlled emissions, enhanced microbial activity, and improved degradation of recalcitrant materials.[50] A pilot scale treatability study evaluated two soil treatment methods for remediating the Brio Refining Superfund Site.[71] At this site, bioremediation was being considered as an alternative to incineration of soils contaminated with chlorinated solvents, aromatic hydrocarbons, and polycyclic aromatic hydrocarbons. Two bioremediation strategies were tested: a land treatment method in a lined bed and a slurry bioreactor. Of the two treatment technologies being tested, the slurry bioreactor had a greater loss of volatile organics due to air-stripping and a greater rate of biodegradation of polynuclear aromatic hydrocarbons.

A slurry reactor, constructed with 208 L drums, was filled with a soil-water slurry and nutrients to obtain a slurry consisting of 30% dry weight solids.[71] The slurry was amended with an inoculum prepared using indigenous microorganisms and treated for 10 days. Air-stripping removed greater than 99% of the volatile organic carbon at a rate that was six times faster than tilling of the treatment bed. Biodegradation of phenanthrene was four times faster in the slurry bioreactor than in the land treatment method. Slurry bioreactor treatment removed 62% of the polynuclear aromatic hydrocarbons within the 10 day treatment period.

The treatment bed used in this comparison study was constructed with a plastic liner, a sand layer, and a drainage system connected to a sump.[71] The treatment bed was enclosed by greenhouses. The bed was filled with soil that was tilled daily to promote oxygen transfer for the microorganisms. Leachates were collected and transferred to a fermenter used to prepare inocula and nutrient solutions. Inoculum and nutrients were sprayed onto the treatment bed. Tilling resulted in air-stripping greater than 99% of volatile organic compounds (VOC) from the soil. These organics were trapped onto carbon. In this land treatment demonstration, 91% of the polynuclear aromatic hydrocarbons were removed during the 94 day treatment period. Evidence that correlated biological activity with high removal of phenanthrene supported the conclusion that semi-volatile compounds were removed by biodegradation rather than air-stripping.

A pilot scale study examined the efficacy of a rotating drum bioreactor (Figure 7) for the treatment of oil-contaminated soil.[18] The reactor was a modified composting system ordinarily used for the treatment of household garbage. The bioreactor was 25 m long and 3.5 m in diameter and rotates to allow the mixing of soil and oil. Soil moisture was maintained by an internal sprinkler system, and nutrients were added to the soil prior to entering the bioreactor. The capacity of the reactor was 50 tons and was tested in batch and semi-continuous modes for periods up to 3 weeks. In batch mode, the oil degrading microflora was established within 3 days, and the majority of biodegradation achieved within the first 10 days. Final concentrations of petroleum in treated soils ranged from approximately 50 to 350 mg/kg. Operating temperatures were about 20°C. Optimum temperatures, approximately 30°C, were not achieved within this reactor. Although "clean" soil (<50 mg/kg) was not achieved, biological treatment did remove the most hazardous components. What remained was predominantly long chain and highly branched alkanes.

3.3 Treatment of Gases

The coupling of biological treatment to air-stripping or vapor phase vacuum extraction of hazardous organic compounds is comparable to the historical use of biological treatment for the control of odorous gases emitted from industrial, agricultural, solid waste, and wastewater treatment facilities.[10,72,73] Applicable to the treatment of gas streams are bioscrubbers, trickling filters, and biofilters.[74] Bioscrubber gas treatment systems utilize freely suspended biomass as in CSTRs, while trickling filters and biofilter systems utilize immobilized biomass with mobile or stationary aqueous phases, respectively.[74]

All gas treatment systems must maintain microbial activity and maximize the mass transfer of gas phase constituents into the liquid phase of the bioreactor, i.e., the volatile organic contaminants, the oxygen for aerobic metabolism, and for the cometabolic degradation of chlorinated solvents by methanotrophic bacteria, the

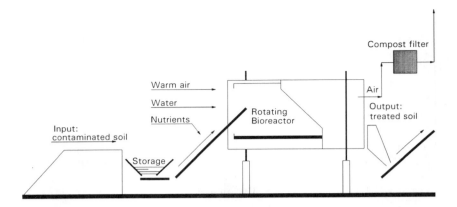

Figure 7. Schematic of a rotating drum bioreactor used for the treatment of contaminated soils. Redrawn from Ger et al.[18]

cosubstrates methane or propane. Nitrogen and phosphorous can be added to trickling filters and bioscrubbers via the liquid phase.

In immobilized cell reactors, the stationary support media should be porous to allow gas flow throughout the bioreactor and provide a high surface area with excellent wetting and sorptive characteristics. For biofilter systems, the solid support frequently serves as the source of inorganic nutrients, and in some cases, the cosubstrate serves for cometabolic degradation. Biofilter support media should provide adequate buffering capacity. Moisture, one of the important process parameters for biofilters, needs to be continually supplied to the biofilter to counteract the drying effect of the gas stream.

Biofilters, used successfully to treat odors as well as toxic inorganic and organic pollutants, have been in widespread use in the European community for the past 20 years.[73] Most biofilters, designed for multi-year operation, have been built as single open beds. Some multiple stage open or closed biofilters are in operation.[73] For the development of a transportable biofilter that is to be used for the treatment of gasoline vapors, Douglass et al.[10] evaluated biofilter packings for flow properties, contaminant retention, contaminant transport and removal, and cost. The air, supplied at the top of vertical columns, provided a constant contaminant concentration at a 100% humidity level. Peat buffered with $CaCO_3$ and Ottawa sand resulted in the greatest removal rates for gasoline and toluene. Contaminant removal rates decreased when contaminant loadings exceeded the maximum biological degradation rates. Packing media such as composts, sand, bark peat, heather, volcanic ash, or mixtures of these materials have also been used in the treatment of VOCs.[73] Tiwaree et al.[75] evaluated different fabrics such as rayon, acryl, polyester, nylon, etc. as carriers of microorganisms in the biological deodorization of dimethyl sulfide.

Immobilized cell bioreactors, such as trickle filters, allow for high surface to volume ratios and high mass transfer rates. Immobilized cell bioreactors can be operated as plug-flow reactors or in recycle mode.[76] Klasson et al. have demonstrated higher efficiency of gas conversion with immobilized cell bioreactors than CSTRs.[77,78] Due to the high cell density and no stirring requirements of packed bed bioreactors, they are potentially more economically feasible to operate. Trickle bed columns allow gas, in the continuous phase, to pass over microbial cells growing on a fixed surface with nutrients passing over the surface. In order to increase reaction rates in these bioreactors the solid supports can be configured to increase surface area and allow for higher cell density. Nutrient flow can be either concurrent or countercurrent to the flow of gas.

Apel et al. evaluated bench scale thin-film bioreactors for the treatment of trichloroethylene and xylene vapor streams.[76] The methanotroph *Methylosinus trichosporium* OB-3b was used for the cometabolic degradation of trichloroethylene using methane as the cosubstrate. A closed-loop bioreactor was operated in batch mode with a recirculating gas stream of 5% methane in air. Gas flow was upward through the column that was packed Plastic Pall rings, which were the solid support for microbial growth. The TCE was added as a vapor to the recirculating air stream.

A similar reactor operated in plug-flow mode demonstrated the degradation of xylene vapors by *Pseudomonas putida*.[76] In this case, xylene served as the carbon source for the growth of this microorganism. A single pass upward flow of xylene/air mixture was countercurrent to a liquid phase of basal salts.

Hollow fiber bioreactors may have some application to the biological destruction of VOCs. The use of hollow fiber bioreactors has been used as a means to increase biological methane production rates from hydrogen and carbon dioxide.[79] The hollow fibers allowed the methanogens attached on the inside of the hollow fibers to have increased contact with liquid media and gases that were added from outside of the fibers.[79] Methanogen growth plugged the hollow fibers with time, but was overcome by back-flushing of the cells out of the reactor with slight positive pressure. Remaining cells acted as the inoculum for the next experiment.

4. SUMMARY

The information presented within this chapter and any one of a number of conference proceedings, technical articles, and textbooks will provide ample evidence that there are in use, being tested, and proposed many approaches for the treatment of hazardous waste and environmental remediation. Each situation is essentially unique with regard to equipment design and operational parameters. The difference is largely due to the chemical and physical characteristics of the hazardous material and the physiological requirements of the microbial system. The physical characteristics determine the methods by which the hazardous materials are brought into contact with the microorganism or enzyme system. The physiological requirements and limitations of the microorganisms determine the operational conditions of the bioreactor and the rate at which the materials are contacted with the microorganism.

When compared bioreactors, the use of farming, composting, and *in situ* remediation for waste treatment and environmental reclamation are relatively simple operations. While improved process control can be obtained with bioreactors, these benefits are offset by the need to transport the contaminated material to the bioreactor and increased capital and operational costs.[80] For many situations, biological processes for hazardous waste treatment and environmental remediation are routine. However, process development is still required for the treatment of recalcitrant or toxic chemicals, chemicals for which cometabolism is the only known mechanism for biological degradation or for materials with an inherent mass transfer limitation.

5. ACKNOWLEDGMENTS

The author wishes to thank Charles E. Turick for comments on the draft of this manuscript and John E. Wey for graphics. This work was supported, in part, by the Department of Energy to the Idaho National Engineering Laboratory under contract No. DE-AC07-76ID01570.

6. REFERENCES

1. Litchfield, C. D. "Practices, Potential, and Pitfalls in the Application of Biotechnology to Environmental Problems," in *Environmental Biotechnology for Waste Treatment*, G. S. Sayler and J. W. Blackburn, Eds. (New York: Plenum Press, 1991), pp. 147–157.
2. Markland Day, S. "Regulatory Considerations," in *Biotechnology for the Treatment of Hazardous Waste* (Chelsea, MI: Lewis Publishers, 1993), pp. 195–221.
3. Hardaway, K. L., M. S. Katterjohn, C. A. Lang, and M. E. Leavitt. "Feasibility and Other Considerations for Use of Bioremediation in Subsurface Areas," in *Environmental Biotechnology for Waste Treatment*, G. S. Sayler and J. W. Blackburn, Eds. (New York: Plenum Press, 1991), pp. 111–125.
4. Flathman, P. E., D. E. Jerger, and L. S. Bottomley. "Remediation of Contaminated Groundwater Using Biological Techniques," *Ground Water Monitoring Rev.* 9:105–119 (1989).
5. Flathman, P. E., D. E. Jerger, P. M. Woodhull, and D. K. Knox. "Remediation of Dichloromethane-Contaminated Groundwater," in *Innovative Hazardous Waste Treatment Technology Series, Volume 3, Biological Processes*, H. M. Freeman and P. R. Sferra, Eds. (Lancaster, PA: Technomic Publishing Co., Inc., 1991), pp. 25–37.
6. Wickramanayake, G. B., H. Nack, and B. R. Allen. "Treatment of Trichloroethylene-Contaminated Groundwater Using Aerobic Bioreactors," in *Innovative Hazardous Waste Treatment Technology Series, Volume 3, Biological Processes*, H. M. Freeman and P. R. Sferra, Eds. (Lancaster, PA: Technomic Publishing Co., Inc., 1991), pp. 169–175.
7. Bartha, R., and I. Bossert. "The Treatment and Disposal of Petroleum Wastes," in *Petroleum Microbiology*, R. M. Atlas, Ed. (New York: Macmillan Publishing Company, 1984), pp. 553–577.
8. Westermeyer, W. E. "Oil Spill Response Capabilities in the United States," *Environ. Sci. Technol.* 25:196–200 (1991).

9. Kelso, D. D., and M. Kendziorek. "Alaska's Response to the *Exxon Valdez* Oil Spill," *Environ. Sci. Technol.* 25:16–23 (1991).
10. Douglass, R. H., J. M. Armstrong, and W. M. Korreck. "Design of a Packed Column Bioreactor for the On-Site Treatment of Air Stripper Off Gas," in *On Site Bioreclamation: Processes for Xenobiotic and Hydrocarbon Treatment*, R. E. Hinchee and R. F. Olfenbuttel, Eds. (Stoneham, MA: Butterworth-Heinemann, 1991), pp. 209–225.
11. Urlings, L. G. C. M., F. Spuy, S. Coffa, and H. B. R. J. van Vree. "Soil Vapour Extraction of Hydrocarbons: In Situ and On-Site Biological Treatment," in *In Situ Bioreclamation: Applications and Investigations for Hydrocarbon and Contaminated Site Remediation*, R. E. Hinchee and R. F. Olfenbuttel, Eds. (Boston: Butterworth-Heinemann, 1991), pp. 321–336.
12. Herrling, B., J. Stamm, and W. Buermann. "Hydraulic Circulation System for In Situ Bioreclamation and/or In Situ Remediation of Strippable Contamination," in *In Situ Bioreclamation: Applications and Investigations for Hydrocarbon and Contaminated Site Remediation*, R. E. Hinchee and R. F. Olfenbuttel, Eds. (Boston: Butterworth-Heinemann, 1991), pp. 173–195.
13. Mackay, D. M., and J. A. Cherry. "Groundwater Contamination: Pump and Treat Remediation," *Environ. Sci. Technol.* 23:630–636 (1989).
14. Canter, L. W., and R. C. Knox. "Physical Control Measures," in *Ground Water Pollution Control* (Chelsea, MI: Lewis Publishers, Inc., 1990), pp. 13–87.
15. *HazTECH News: The Newsletter of Hazardous Waste Treatment Technology* 8(3):17 (February 11, 1993).
16. Bartha, R. "Biotechnology of Petroleum Pollutant Biodegradation," *Microb. Ecol.* 12:155–172 (1986).
17. Dibble, J. T., and R. Bartha. "Rehabilitation of Oil-Inundated Agricultural Land: A Case History," *Soil Sci.* 140:75–77 (1979).
18. Ger, P., M. van den Munckhof, and M. F. X. Veul. "Production-Scale Trials on the Decontamination of Oil-Polluted Soil in a Rotating Bioreactor at Field Capacity," in *On Site Bioreclamation: Processes for Xenobiotic and Hydrocarbon Treatment*, R. E. Hinchee and R. F. Olfenbuttel, Eds. (Boston: Butterworth-Heinemann, 1991), pp. 443–451.
19. Compeau, G. C., W. D. Mahaffey, and L. Patras. "Full-Scale Bioremediation of Contaminated Soil and Water," in *Environmental Biotechnology for Waste Treatment*, G. S. Sayler and J. W. Blackburn, Eds. (New York: Plenum Press, 1991), pp. 91–109.
20. Hinchee, R. E., R. N. Miller, and R. Ryan Dupont. "Enhanced Biodegradation of Petroleum Hydrocarbons: An Air-Based In Situ Process," in *Innovative Hazardous Waste Treatment Technology Series, Volume 3, Biological Processes*, H. M. Freeman and P. R. Sferra, Eds. (Lancaster, PA: Technomic Publishing Co., Inc., 1991), pp. 177–183.
21. Hutchins, S. R., and J. T. Wilson. "Laboratory and Field Studies on BTEX Biodegradation in a Fuel-Contaminated Aquifer Under Denitrifying Conditions," in *In Situ Bioreclamation: Applications and Investigations for Hydrocarbon and Contaminated Site Remediation*, R. E. Hinchee and R. F. Olfenbuttel, Eds. (Boston: Butterworth-Heinemann, 1991), pp. 157–172.
22. Rainwater, K., and R. J. Scholze, Jr. "In Situ Biodegradation for Treatment of Contaminated Soil and Groundwater," in *Innovative Hazardous Waste Treatment Technology Series, Volume 3, Biological Processes*, H. M. Freeman and P. R. Sferra, Eds. (Lancaster, PA: Technomic Publishing Co., Inc., 1991), pp. 107–121.

23. McCarty, P. L., L. Semprini, M. E. Dolan, T. C. Harmon, C. Tiedeman, and S. M. Gorelick. "In Situ Methanotrophic Bioremediation for Contaminated Groundwater at St. Joseph, Michigan," in *On Site Bioreclamation: Processes for Xenobiotic and Hydrocarbon Treatment*, R. E. Hinchee and R. F. Olfenbuttel, Eds. (Stoneham, MA: Butterworth-Heinemann, 1991), pp. 16–40.

24. Semprini, L., G. D. Hopkins, P. L. McCarty, and P. V. Roberts. "In Situ Transformation of Carbon Tetrachloride and Other Halogenated Compounds Resulting from Biostimulation Under Anoxic Conditions," *Environ. Sci. Technol.* 26:2454–2461 (1992).

25. Semprini, L., P. V. Roberts, G. D. Hopkins, and P. L. McCarty. "A Field Evaluation of In-Situ Biodegradation of Chlorinated Ethenes. II. Results of Biostimulation and Biotransformation Experiments," *Ground Water* 28:715–727 (1990).

26. Bartha, R. "Biotechnology of Petroleum Pollutant Biodegradation," *Microb. Ecol.* 12:155–163 (1986).

27. Vogel, T. M., C. S. Criddle, and P. L. McCarty. "Transformations of Halogenated Aliphatic Compounds," *Environ. Sci. Technol.* 21:722–736 (1987).

28. Grady, C. P. L., Jr., and H. C. Lim. "Classification of Biochemical Operations," in *Biological Wastewater Treatment* (New York: Marcel Dekker, Inc., 1980), pp. 3–14.

29. Eckenfelder, W. W., Jr. "Biological Wastewater-Treatment Processes," in *Industrial Water Pollution Control*, 2nd ed. (New York: McGraw Hill Book Company, 1989), pp. 189–262.

30. Sutton, P. M. "Engineered Systems for Biotreatment-Adsorption of Hazardous Wastes," in *International Conference on Physical and Biological Detoxification of Hazardous Wastes*, Y. C. Wu, Ed. (Lancaster, PA: Technomic Publishing Co., Inc., 1989), pp.53–74.

31. Chang, H. N., and M. Moo-Young. "Analysis of Oxygen Transport in Immobilized Whole Cells," in *Bioreactor Immobilized Enzymes and Cells: Fundamentals and Applications*, M. Moo-Young, Ed. (New York: Elsevier, 1988), pp. 33–51.

32. Savage, G. M., L. F. Diaz, and C. G. Golueke. "Disposing of Organic Hazardous Wastes by Composting," *BioCycle* 26:31–34 (1985).

33. Hart, S. A. "Composting Potentials for Hazardous Waste Management," in *Innovative Hazardous Waste Treatment Technology Series, Volume 3, Biological Processes*, H. M. Freeman and P. R. Sferra, Eds. (Lancaster, PA: Technomic Publishing Co., Inc., 1991), pp. 7–17.

34. Atlas, R. M., and R. Bartha "Stimulated Biodegradation of Oil Slicks Using Oleophilic Fertilizers," *Environ. Sci. Technol.* 7:538–541 (1973).

35. Lee, K., and E. Levy. "Biodegradation of Petroleum in the Marine Environment: Limiting Factors and Methods of Enhancement," Canadian Technical Report of Fisheries and Aquatic Sciences, No. 1442 (1986).

36. Glaser, J. A. "Nutrient-Enhanced Bioremediation of Oil-Contaminated Shoreline: The Valdez Experience," in *On Site Bioreclamation: Processes for Xenobiotic and Hydrocarbon Treatment*, R. E. Hinchee and R. F. Olfenbuttel, Eds. (Stoneham, MA: Butterworth-Heinemann, 1991), pp. 366–384.

37. Lenzo, F., and D. G. Ward, Jr. "Aerobic Biodegradation of Low Levels of Dissolved Organic Compounds in Groundwater: Research and Field Data," in *On Site Bioreclamation: Processes for Xenobiotic and Hydrocarbon Treatment*, R. E. Hinchee and R. F. Olfenbuttel, Eds. (Stoneham, MA: Butterworth-Heinemann, 1991), pp. 422–428.

38. Folsom, B. R., P. J. Chapman, and P. H. Pritchard. "Phenol and Trichloroethylene Degradation by *Pseudomonas cepacia* G4: Kinetics and Interactions Between Substrates," *Appl. Environ. Microbiol.* 56:1279–1285 (1990).

39. Little, C. D., A. V. Palumbo, S. E. Herbes, M. E. Lindstrom, R. L. Tyndall, and P. J. Gilmer. "Trichloroethylene Biodegradation by a Methane Oxidizing Bacterium," *Appl. Environ. Microbiol.* 54:951–956 (1988).

40. Roberts, P. V., L. Semprini, G. D. Hopkins, D. Grbić-Galić, P. L. McCarty, and M. Reinhardt. "*In situ* Aquifer Restoration of Chlorinated Aliphatics by Methanotrophic Bacteria," U.S. Environmental Protection Agency, Report EPA/600/2/89/033 (1989).

41. Phelps, T. J., J. J. Niedielski, R. M. Schram, S. E. Herbes, and D. C. White. "Biodegradation of Trichloroethylene in Continuous-Recycle Expanded-Bed Bioreactors," *Appl. Environ. Microbiol.* 56:1702–1709 (1990).

42. Speitel, G. E., and J. M. Leonard. "A Sequencing Biofilm Reactor for the Treatment of Chlorinated Solvents Using Methanotrophs," *Water Environ. Res.* 64: 712–719 (1992).

43. Oldenhuis, R., R. L. J. M. Vink, D. B. Janssen, and B. Witholt. "Degradation of Chlorinated Aliphatic Hydrocarbons by *Methylosinus trichosporium* OB3b Expressing Soluble Methane Monooxygenase," *Appl. Environ. Microbiol.* 55:2819–2826 (1989).

44. Tsien, H. C., G. A. Brusseau, R. S. Hanson, and L. P. Wackett. "Biodegradation of Trichloroethylene by *Methylosinus trichosporium* OB-3b," *Appl. Environ. Microbiol.* 55:3155–3161 (1989).

45. *HazTECH News: The Newsletter of Hazardous Waste Treatment Technology* 7(26):200 (December 31, 1992).

46. Tabak H. H., J. A. Glaser, S. Strohofer, M. J. Kupferle, P. Scarpino, and M. W. Tabor. "Characterization and Optimization of Treatment of Organic Wastes and Toxic Organic Compounds by a Lignolytic White Rot Fungus in Bench-Scale Bioreactors," in *On Site Bioreclamation: Processes for Xenobiotic and Hydrocarbon Treatment*, R. E. Hinchee and R. F. Olfenbuttel, Eds. (Stoneham, MA: Butterworth-Heinemann, 1991), pp. 341–365.

47. Zitomer, D. H., and R. E. Speece. "Sequential Environments for Enhanced Biotransformation of Aqueous Contaminants," *Environ. Sci. Technol.* 27:227–244 (1993).

48. Fathepure, B. Z., and T. M. Vogel. "Complete Degradation of Polychlorinated Hydrocarbons by a Two-Stage Biofilm Reactor," *Appl. Environ. Microbiol.* 57:3418–3422 (1991).

49. Mueller, J. G., S. E. Lantz, D. Ross, R. J. Colvin, D. P. Middaugh, and P. H. Pritchard. "Strategy Using Bioreactors and Specially Selected Microorganisms for Bioremediation of Groundwater Contaminated with Creosote and Pentachlorophenol," *Environ. Sci. Technol.* 27:691–698 (1993).

50. Stegmann, R., S. Lotter, and J. Heerenklage. "Biological Treatment of Oil-Contaminated Soils in Bioreactors," in *On Site Bioreclamation: Processes for Xenobiotic and Hydrocarbon Treatment*, R. E. Hinchee and R. F. Olfenbuttel, Eds. (Stoneham, MA: Butterworth-Heinemann, 1991), pp. 188–208.

51. Singer, M. E., and W. R. Finnerty. "Microbial Metabolism of Straight-Chain and Branched Alkanes," in *Petroleum Microbiology*, R. M. Atlas, Ed. (New York: Macmillan Publishing Co., 1984), pp. 1–59.

52. Cerniglia, C. E. "Microbial Transformation of Aromatic Hydrocarbons," in *Petroleum Microbiology*, R. M. Atlas, Ed. (New York: Macmillan Publishing Co., 1984), pp. 99–128.

53. Stoner, D. L. "Hazardous Organic Waste Amenable to Microbial Transformation," in *Biotechnology for the Treatment of Hazardous Waste* (Chelsea, MI: Lewis Publishers, 1993), pp. 1–25.

54. *HazTECH News: The Newsletter of Hazardous Waste Treatment Technology* 7(26):200 (December 31, 1992).

55. Huling, S. G., B. E. Bledsoe, and M. V. White. "The Feasibility of Utilizing Hydrogen Peroxide as a Source of Oxygen in Bioremediation," in *In Situ Bioreclamation: Applications and Investigations for Hydrocarbon and Contaminated Site Remediation*, R. E. Hinchee and R. F. Olfenbuttel, Eds. (Boston: Butterworth-Heinemann, 1991), pp 83–102.

56. Armstrong, J. M., and J. T. Wilson. "National Demonstration Site: In-Situ Bioremediation of Contaminated Aquifer," in *Biotechnology Applications in Hazardous Waste Treatment*, G. Lewandowski, P. Armenante, and B. Baltzis, Eds. (New York: United Engineering Trustees, Inc., 1989), pp. 281–288.

57. Lee, M. D., and R. L. Raymond, Sr. "Case History of the Application of Hydrogen Peroxide as an Oxygen Source for In Situ Bioreclamation," in *In Situ Bioreclamation: Applications and Investigations for Hydrocarbon and Contaminated Site Remediation*, R. E. Hinchee and R. F. Olfenbuttel, Eds. (Boston: Butterworth-Heinemann, 1991), pp 429–436.

58. Major, D. W., E. W. Hodgins, and B. J. Butler. "Field and Laboratory Evidence of In Situ Biotransformation of Tetrachloroethene to Ethene and Ethane at a Chemical Transfer Facility in North Toronto," in *On Site Bioreclamation: Processes for Xenobiotic and Hydrocarbon Treatment*, R. E. Hinchee and R. F. Olfenbuttel, Eds. (Stoneham, MA: Butterworth-Heinemann, 1991), pp. 147–171.

59. Novick, N. J., and M. Alexander. "Cometabolism of Low Concentrations of Propachlor, Alachlor, and Cycloate in Sewage and Lake Water," *Appl. Environ. Microbiol.* 49:737–743 (1985).

60. Alkajjar, B. J., G. V. Simsiman, and G. Chester. "Fate and Transport of Alachlor, Metachlor, and Atrazine in Large Columns," *Water. Sci. Technol.* 22:87–94 (1990).

61. Hogan, J. A., G. R. Toffoli, F. C. Miller, J. V. Hunter, and M. S. Finstein. "Composting Physical Model Demonstration: Mass Balance of Hydrocarbons and PCBs," in *Physiochemical and Biological Detoxification of Hazardous Wastes, Volume 2. Proceedings of an International Conference May 3–5, 1988 Atlantic City, New Jersey* (Lancaster, PA: Technomic Publishing Co., Inc., 1989), pp. 742–757.

62. Dibble, J. T., and R. Bartha. "Effect of Environmental Parameters on the Biodegradation of Oil Sludge," *Appl. Environ. Microbiol.* 37:729–739 (1979).

63. Bossert, I., W. M. Kachel, and R. Bartha. "Fate of Hydrocarbons During Oily Sludge Disposal in Soil," *Appl. Environ. Microbiol.* 47:763–767 (1984).

64. Lamar, R. T., and D. M. Dietrich. "In Situ Depletion of Pentachlorophenol from Contaminated Soil by *Phanaerochaete* spp.," *Appl. Environ. Microbiol.* 56:393–3100 (1990).

65. Bumpus, J. A., and S. Aust. "Biodegradation of Environmental Pollutants by the White Rot Fungus, *Phanerochaete chrysosporium*: Involvement of the Lignin Degrading System," *BioEssays* 6:166–170 (1987).

66. Bumpus, J. A., G. Mileski, B. Brock, W. Ashbaugh, and S. D. Aust. "Biological Oxidations of Organic Compounds by Enzymes from a White Rot Fungus," in *Innovative Hazardous Waste Treatment Technology Series, Volume 3, Biological Processes*, H. M. Freeman and P. R. Sferra, Eds. (Lancaster, PA: Technomic Publishing Co., Inc., 1991), pp. 169–175.

67. Field, J. A., E. de Jong, G. Fiejo Costa, and J. A. M. de Bont. "Biodegradation of Polycyclic Aromatic Hydrocarbons by New Isolates of White Rot Fungi," *Appl. Environ. Microbiol.* 58:2219–2226 (1992).
68. Morgan, P., S. T. Lewis, and R. J. Watkinson. "Comparison of Abilities of White-Rot Fungi to Mineralize Selected Xenobiotic Compounds," *Appl. Microbiol. Biotechnol.* 34:693–696 (1991).
69. Aust, S. D. "Degradation of Environmental Pollutants by *Phanerochaete chrysosporium*," *Microb. Ecol.* 20:197–209 (1990).
70. Qiu, X. J., and M. J. McFarland. "Bound Residue Formation in PAH Contaminated Soil Composting Using *Phanerochaete chrysosporium*," *J. Hazardous Wastes and Hazardous Materials* 8:115–126 (1991).
71. Yare, B. S. "A Comparison of Soil-Phase and Slurry-Phase Bioremediation of PNA-Contaminated Soils," in *On Site Bioreclamation: Processes for Xenobiotic and Hydrocarbon Treatment*, R. E. Hinchee and R. F. Olfenbuttel, Eds. (Stoneham, MA: Butterworth-Heinemann, 1991), pp. 173–187.
72. Williams, T. O., and F. C. Miller, "Biofilters and Facility Operations," *BioCycle* 33:75–79 (1992).
73. Leson, G., and A. M. Winer. "Biofiltration: An Innovative Air Pollution Control Technology for VOC Emissions," *J. Air Waste Manage. Assoc.* 41:1045–1054 (1991).
74. Diks, R. M. M., and S. P. P. Ottengraf. "A Biological Treatment System for the Purification of Waste Gases Containing Xenobiotic Compounds," in *On Site Bioreclamation: Processes for Xenobiotic and Hydrocarbon Treatment*, R. E. Hinchee and R. F. Olfenbuttel, Eds. (Stoneham, MA: Butterworth-Heinemann, 1991), pp. 452–463.
75. Tiwaree, R. S., K. S. Cho, M. Hirai, and M. Shoda. "Biological Deodorization of Dimethyl Sulfide Using Different Fabrics as the Carriers of Microorganisms," *Appl. Biochem. Biotechnol.* 32:135–148 (1992).
76. Apel, W. A., P. R. Dugan, M. R. Wiebe, E. G. Johnson, J. H. Wolfram, and R. D. Rogers. "Bioprocessing of Environmentally Significant Gases and Vapors with Gas-Phase Bioreactors," in *Emerging Technologies in Hazardous Waste Management III*, D. W. Tedder and F. G. Pohland, Eds. (Washington, DC: American Chemical Society, 1993), pp. 411–428.
77. Klasson, K. T., C. M. D. Ackerson, E. C. Clausen, and J. L. Gaddy. "Mass-Transport in Bioreactors for Coal Synthesis Gas Fermentation," *Abstr. Pap. Am. Chemical Soc., Fuel* 204:125 (1992).
78. Klasson, K. T., C. M. D. Ackerson, E. C. Clausen, and J. L. Gaddy. "Biological Conversion of Synthesis Gas into Fuels," *Int. J. Hydrogen Energy*, 17:281–288 (1992).
79. Jee, H. S., N. Nishio, and S. Nagai, "CH_4 Production from H_2 and CO_2 by *Methanobacterium thermoautotrophicum* Cells Fixed on Hollow Fibers," *Biotechnol. Lett.* 10:243–248 (1988).
80. Yano, T., K. Aoki, and S. Nagai. "Kinetics of CH_4 Production from H_2 and CO_2 in a Hollow Fiber Reactor by Plug Flow Reaction Model," *J. Fermentation and Bioengineer* 71:203–205 (1991).

CHAPTER 4

Biotechnological Treatment of Liquid and Solid Inorganic Wastes

W. D. Gould and R. G. L. McCready

1. INTRODUCTION

Many inorganic wastes are difficult and expensive to treat by current chemical and physical treatment technologies; an alternative may be found in biologically based technologies. Biotechnology based procedures for the treatment of inorganic wastes are in different states of development varying from laboratory experiments to pilot plant tests. Inorganic pollutants can be found in liquid and solid industrial effluents and are produced by mining, metal refining, manufacturing industries, and also by agricultural activities. This chapter discusses treatment of inorganic wastes with respect to the presently accepted technology, the state of biotechnology in treatment of each specific waste, and future possibilities of biotechnology for the amelioration of inorganic pollutants.

0-87371-613-2/94/$0.00+$.50
© 1994 by Lewis Publishers

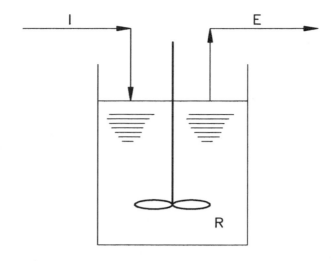

Figure 1.   Continuously stirred tank reactor (I: influent, E: effluent, R: reaction zone).

2. BIOREACTORS USED IN LIQUID WASTE TREATMENT

A variety of reactor types can be used to contact a liquid effluent with a biological agent. Reactors can be used to contact an effluent with either a biological material such as dead immobilized fungi or an active microbial culture. Reactors can be operated in either batch or continuous mode. In batch mode, the microbial inoculant and effluent are introduced into the reactor, and microbial growth with the concurrent reactions is allowed to proceed until the reaction is complete before the effluent is discharged. In the continuous mode, treated effluent is withdrawn at the same rate at which it is added. The contents of a reactor operated in the continuous mode are at steady-state; consequently, they do not vary with time. The critical parameter in the operation of a continuous reactor is the flow rate.

The continuously stirred tank reactor (CSTR) is essentially a cylindrical vessel in which the contents are well mixed (Figure 1) and has provisions for addition of feed, removal of product, mixing, and aeration. The CSTR can be used in either continuous or batch mode. The air-lift fermenter consists of a column with an internal draft-tube (Figure 2). Compressed air is sparged into the base of the draft-tube, which causes the air-entrapped liquid to rise in the draft-tube and return through the annular space on the sides. The air-induced circulation provides both mixing and oxygen transfer. Usually, air-lift fermenters are operated in the batch mode.

The fixed film class of reactors consists of a solid support to which the biomass adheres. The biomass, thus immobilized, is not subject to being washed out, which can occur in other types of reactors. The packed tower, the rotating biological contactor (RBC), and the fluidized bed are examples of fixed film reactors. The

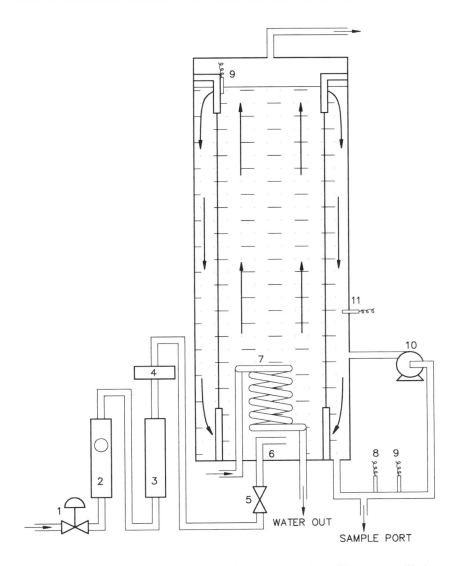

Figure 2. Air-lift fermentor consisting of: (1) air pressure regulator, (2) rotameter, (3) glass-wool air filter, (4) millipore air filter, (5) check valve, (6) air sparger, (7) heating and cooling coil, (8) pH electrode, (9) oxygen probe, (10) centrifugal pump, and (11) thermister probe.

RBC is essentially a rotating shaft supporting a series of disks partially immersed in the effluent that is to be treated (Figure 3). Microorganisms colonizing the surface of the disks modify, degrade, and/or absorb pollutants in the effluent streams. The effluent enters the first chamber of the RBC and follows a serpentine path through the chambers of the RBC, which allows maximum contact of the

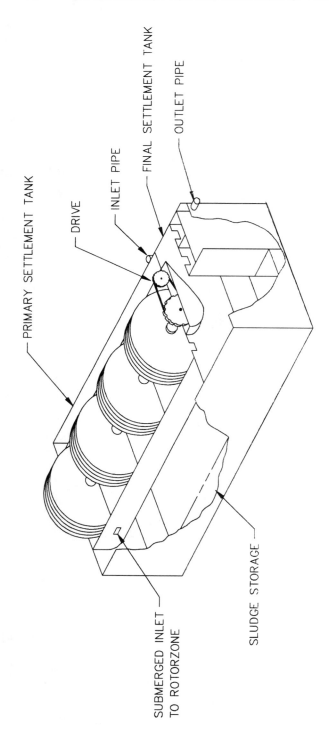

Figure 3. Rotating biological contactor.

solution with the biomass. The RBC is generally operated in the continuous mode. The RBC offers a number of advantages: (1) low energy requirements, (2) simplicity of operation, (3) low maintenance requirements, (4) high treatment efficiency, (5) excellent oxygen transfer, and (6) resistance to shock loads. The disadvantages of the RBC are the somewhat higher biochemical oxygen demand (BOD) and suspended solids content of the final effluent compared to that of other types of reactors.

The packed tower can be operated in either the upward or downward flow modes. The trickling filter is a packed tower reactor that consists of a bed of stones or sand, which supports the biomass, and the effluent is allowed to trickle over the solid support (Figure 4). The packed tower can also be operated in the upward flow mode, which will allow the development of anoxic conditions within the reactor. The fluidized bed reactor is a packed tower operated in the upward flow mode such that the solid support is fluidized. The packed tower is only operated in a continuous mode.

3. LIQUID WASTES

3.1 Metals and Radionuclides

Metal-containing liquid wastes produced by industrial activities vary from small volumes of solution containing high concentrations of metal ions to large volumes containing dilute concentrations of metal ions. Examples include waste-waters emanating from film processing plants,[1] tanneries, electroplating industries, and mining, smelting, and refining operations,[2] radionuclides released by uranium mining,[3] and nuclear fuel reprocessing plants. Conventional techniques for the removal of metal ions, such as ion exchange and precipitation, lack specificity and are ineffective at removing low concentrations of metal ions. Biosorption onto bacteria,[4,5] fungi,[6,7] yeasts,[8,9] actinomycetes,[10] algae,[11–13] plant tissue,[14] activated sludge,[15] and various biopolymers[16,17] have been used for the removal of metal ions from solution.

Accumulation of metals and radionuclides can occur by enzymatic reactions such as active transport or by nonenzymatic reactions such as sorption onto cellular components. Active transport, the transfer of materials across the cell membrane, can be metal specific. For example, plants, bacteria, and fungi have evolved high affinity iron uptake systems, which are induced by low iron concentrations and repressed by high iron levels.[18] However, because active transport is an energy-requiring process, live, respiring biomass is necessary.

Living or nonliving microbial cells can sequester metal ions by cation exchange, chelation,[19] or adsorption. Chelation or ion exchange can be mediated by various functional groups present in most biological polymers: carboxyl, sulfate, sulfonic acid, phosphate, hydroxyl, sulfhydryl, amino, lactam (–NH), imidazole, and imino functional groups.[16] Metals can bind to extracellular polymers, the cell wall, or various components in the cytoplasm, although most binding occurs in the

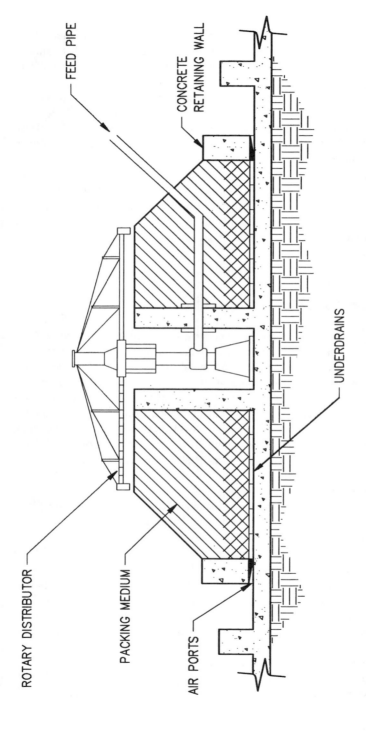

Figure 4. Trickling filter reactor.

cell wall or cell envelope region. More than one mechanism can be involved in sequestering metals, and for that reason, the degree of specificity of the bioprocess can be controlled somewhat. For instance, the biosorption of uranium by *Rhizopus arrhizus* is mediated by the chitin in the cell wall and occurs in two stages.[20,21] Amino groups of the cell wall chitin act as uranium coordination sites, which then function as nucleation centers for the deposition of uranyl hydroxide. In contrast, the binding of copper (II) by *Rhizopus arrhizus* occurs via coordination of the copper with one nitrogen-containing ligand and with oxygen atoms filling the remainder of the coordination positions.[22] Various biopolymers produced by living cells contain a wide range of functional groups that can bind metals.

Because metal uptake by active mechanisms requires respiring biomass and accounts for a small percentage of the metal sequestered by most microorganisms, the use of dead biomass for biosorption is the most practical alternative for large scale metal removal. Dead biomass has other advantages as well: (1) toxicity of the metals will have no effect on the metal binding capacity; (2) the biomass can be grown under optimum conditions prior to use; and (3) the biomass can be immobilized to provide maximum physical stability during operation. The metal binding capacity and specificity of biomass can be altered both by the growth conditions used to obtain the biomass, selection of the biomass, and the subsequent physical treatment of the biomass.[23] Pighi et al.[24] tested 32 soil fungi for their metal binding capacity and found that all but one of them selectively bound silver ions from solution. A strong preference for silver over copper, lead, nickel, and cadmium was observed. The selectivity shown by biomass for certain metal ions could be used for the removal of economically valuable or toxic metals from solution in the presence of higher concentrations of other metals. The functional groups in the cell walls of the biomass can be chemically derivatized to produce biomass with enhanced specificity and metal binding capacity.[25] Recent studies by Costerton[26] have indicated that bacterial isolates from a nickel mine drainage system have exceedingly high loading capacities for nickel. These cells do not take up iron, but will accumulate lesser amounts of cobalt. Nickel adsorption is maximal at pH 7.0, but a substantial increase in the Ni loading (per g of dry biomass) was observed after a brief exposure of the cells to pH 4.0 prior to the metal adsorption at pH 7.0. Further studies are required to confirm these preliminary findings and to assess the potential use of these organisms to remove nickel from mine effluents.

Morper[27] reported on the use of anaerobic sewage sludge in an upflow anaerobic sludge bed to treat metal laden solutions. During the treatment of polluted wastewater from brandy production (COD 35,000 mg/L with 20 mg Cu/L), the sludge in the lower zone of the reactor contained 9000 mg/L of copper. The major mechanism of metal removal is assumed to be microbial reduction of sulfate to sulfide with the subsequent precipitation of copper sulfide. More recent studies have shown that the sewage sludge system is also efficient in removing precious metals from solution. On treating a silver-containing effluent, the sludge adsorbed up to 54 mg Ag/g (dry wt) of sludge, and for gold, the maximum loading was about 43 mg Au/g (dry wt) of sludge. The metal values can be recovered from the

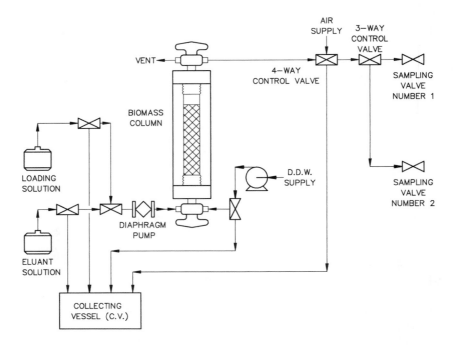

Figure 5. Column containing immobilized biomass for the removal of uranium from dilute solutions.

sludge by incineration followed by dissolution of the metals from the ash using aqua regia.

Low mechanical strength and small particle size are the two major problems encountered with native biomass.[6,28] Various techniques for pelletization or immobilization of biomass have been developed.[29] Biomass can be immobilized by adsorption onto a solid carrier, such as sand,[9] incorporated in gels or polymeric matrices, and bound by various covalent cross-linking agents such as acrylamide.[29] The physical entrapment of biomass in reticulated foam[30] and the pore spaces of polysulfone beads[31] has also been used to immobilize biomass. Although a variety of techniques can be used for contacting the biomass with the wastewater, the most commonly used systems employ the ion exchange configuration (Figure 5).[6]

The removal of radionuclides from waste streams is necessary to prevent the release of radioactivity to the environment and to recover potentially valuable resources. Thorium, which can be used in nuclear fuel, is frequently discharged in effluents during uranium mining. Nuclear fuel reprocessing plants produce effluent streams containing up to 20 different radionuclides.[32] The conventional methods to concentrate and remove radionuclides from dilute effluents are coprecipitation with barium, ion exchange, or solvent extraction. For example, radium is currently being removed from uranium mine effluents by coprecipitation as a barium-radium sulfate sludge by the addition of $BaCl_2$ and limestone to the sulfate-rich effluent. However, there is potential for the redissolution of the radium if freshwater contacts the sludge.

Biosorption is an alternative technique for the accumulation of radionuclides. Tsezos and Keller[33] found that *Penicillium chrysogenum* biomass had an adsorptive capacity for radium that was an order of magnitude higher than conventional materials such as ion exchange resins and activated carbon. Neutral to alkaline pH values were required for biosorption to be effective. Return sludge from a municipal wastewater treatment plant has also been used for the selective removal of radium-226 from uranium mining and milling effluent.[34,35] Dried sludge, which has been immobilized using polymeric cross-linking agents and then pelletized, was used in a batch reactor for radium uptake studies. The radium could be eluted from the immobilized sludge, although some loss of biosorption capacity occurs after each loading and elution cycle.[35] The immobilized sludge could only be reused three times; however, the specificity and high loading capacity of the material compensated for the other disadvantages.

3.2 Acid Mine Drainage

The processing of base metal ores results in the production of large amounts of waste metal sulfides such as pyrite, which are disposed of in tailings ponds. Biological oxidation of these metal sulfides results in the production of sulfuric acid, ferric sulfate, and various other metal sulfates. Consequently, the drainage from these tailing ponds is acidic and may contain very high levels of soluble metals. The conventional treatment for these acidic, metal-laden wastes is neutralization and metal precipitation with calcium hydroxide. However, the use of lime is expensive and must be continued for as long as the tailings produce acidic drainage. A self-sustaining biological treatment system would be far superior to liming.

When mildly acidic drainage enters a swamp or bog, the final effluent leaving the system has a higher pH and lower metal concentrations than the influent. However, very acidic drainage is often not improved by running through natural wetlands. Presently, a number of studies are in progress to examine the mechanisms by which this occurs with a view to optimizing the process and constructing self-sustaining ecosystems for the amelioration of acidic drainage.[36] It has been postulated that the increase in pH observed in wetlands can be attributed to bacterial sulfate reduction and proteolysis.[37–40] Metals can be precipitated as metal sulfides or removed by adsorption or chelation by organic matter. Degradation of organic matter in wetlands will lead to the establishment of anaerobic conditions and generate nutrients for the sulfate-reducing bacteria. Growth of aquatic plants in the wetlands will provide a continuous source of organic matter. In 1973, Gale et al.[41] reported that aquatic plants exposed to metal ions associated with the effluents of lead mines in the Missouri Lead Belt accumulated high concentrations of heavy metals. Later, Gale and Wixson[42] reported that the loading capacity of several aquatic algae ranged from 505 to 836 ppm of lead, 510 to 695 ppm of zinc,

39 to 46 ppm of copper, and 10,353 to 14,287 ppm of manganese when exposed to a lead smelter effluent.

Kalin and Wheeler[36] reported the successful use of "biological polishing" with macrophytic algae in the effluents of a base metal mine in Newfoundland. After implementation, the zinc concentrations in the Decant Pond have been reduced to less than 5 mg/L, which is much lower than the levels achievable with liming alone. Boojum Research has also implemented the use of "Chara" in the biological polishing of alkaline mine effluents and has shown that this macrophyte will accumulate significant quantities of uranium, radium 226, nickel, and copper. Their efforts are now concentrated on the development of biological polishing systems for acidic mine drainage.

P. Lane and Associates[43] have shown that plant and algal species indigenous to an abandoned gold mill tailings accumulate as much as 10,000 ppm of arsenic and 16 ppm of mercury in their tissues. The moss *Campylium stellata* accumulated the highest levels of metals: 4000 ppm of arsenic and 7.7 ppm of mercury. Of the terrestial plants, horsetails and cattails accumulated the largest quantities of As and Hg, but growth was suppressed as the metals accumulated in the plant tissue.

Another approach to prevent the formation of acid drainage from small amounts of acid-generating material is to cover the surface with an impermeable cap. The sealed surface prevents the infiltration of water and oxygen, both of which are necessary for the generation of acidic drainage. One successful project using this technique was established at the Halifax International Airport in Canada. The airport is located on a band of highly mineralized pyritic slate bedrock. Approximately 225,000 cubic meters of waste rock from the construction of a taxiway were deposited in an area covering 7 hectares.[44] Within a short period of time, the rock was generating an acidic effluent containing high concentrations of soluble metals. The waste rock was capped with a 1 cm layer of salt, a 75 cm layer of compacted clay, 15 cm of topsoil, and seeded to grass. The salt layer, which is in contact with both the clay cap and the waste rock, inhibits the acid-generating microorganism *Thiobacillus ferrooxidans* and renders the clay less permeable. Previous to capping, approximately 250 L/min of effluent at pH 3.2 containing 400 mg iron per liter was initially being generated by the waste rock. One year after installation of the clay cap, the quantity of effluent had decreased to 50 L/min with an iron content of 80 mg/L and aluminum content of 25 mg/L. Subsequently, further improvements in water quality have been observed, and the cost of water treatment has been reduced by 70%.

3.3 Selenium

Most selenium compounds are very toxic even though selenium is also an essential trace nutrient for microorganisms, plants, and animals including human beings.[45,46] The U.S. Environmental Protection Agency (USEPA) has set a limit for selenium concentrations in drinking water of 10 ppb[47] and a limit of 1.0 ppm for the discharge of wastewater.[48] Selenium generally occurs in conjunction with sulfur or sulfides. Selenium is produced as a by-product in precious metal refining operations.[47] Selenium also occurs in groundwater,[49] effluents from copper refineries,[50] water used to scrub roaster gases in zinc sulfide smelters,[51] wastewater from oil refineries,[48] and electric power plants.[52]

Chemical techniques for the removal of selenium from effluent streams include precipitation with lime,[53] adsorption on activated alumina,[54] reverse osmosis,[53] ion exchange,[53,55] and chemical reduction.[51] All of these techniques have some disadvantages such as incomplete removal, high cost, and poor selectivity. In addition, chemical procedures are less effective in the removal from solution of the selenate ion than the selenite ion.

Microorganisms can mediate transformations that can alter the solubility, volatility, or availability of selenium compounds. Selenium can be incorporated into specific enzymes or into cellular proteins in place of sulfur and reduced to volatile organo-selenium compounds such as dimethyl selenide.[45,46,56,57] Selenium oxyanions can also be reduced to elemental selenium by microorganisms and thus removed from solution. Although microorganisms capable of reducing the selenate ion have been isolated,[49,58] selenite is more easily reduced than is selenate.[45] *Salmonella heidelberg*,[59] *Escherichia coli*,[60,61] yeast,[62] and various rumen bacteria are among the microorganisms shown to reduce the selenite ion to elemental selenium. Due to unfavorable thermodynamics for the reduction of Se^o to Se^{2-}, elemental selenium is often the final product for the biological reduction of selenium oxyanions. The biochemical pathway for the dissimilatory reduction of the selenium oxyanions is not well understood. In most microorganisms studied to date, selenium reduction has been shown to be different from the dissimilatory sulfate reduction.[63–65] Therefore, competitive inhibition of selenium uptake and reduction is not anticipated to be a problem.

Laboratory studies utilizing packed bed reactors, plug flow reactors, and RBCs have been successful in removing selenate from mining process wastewater and agricultural drainage water.[49,66] Both laboratory and pilot plant studies have been carried out using an RBC containing *E. coli* for the removal of selenite from weak acid smelter effluent.[2] Sewage effluent combined with the wastewater can be used as the source of *E. coli* and carbon. However, *E. coli* will only reduce the selenite ion. Elemental selenium accumulates in the microbial biomass during the dissimilatory reduction of selenium. The resulting sludge can be processed, and the selenium can be recovered as a valuable product. Selenium is used for a number of applications such as photoreceptors, electronic glass decolorization, and pigments.[47]

Gerhard et al.[67] developed a pilot plant scale system for the removal of both nitrate and selenate from agricultural drainage. The first stage of this system employs a pond containing algae to remove some of the nitrate and provide an inexpensive carbon and energy source for the selenate and nitrate-reducing bacteria. The second stage consisted of an anoxic pond that removed most of the nitrate and reduced the selenate primarily to selenite. It was necessary to add ferric chloride to the effluent to remove most of the soluble selenium.

3.4 Cyanide

Cyanide-containing wastes are produced by metal plating operations, gold mills, steel mills, and petrochemical and synthetic fuel plants. Both chemical and biological techniques have been used for the treatment of cyanide-containing wastes. The best characterized technique for cyanide destruction is alkaline chlorination by means of chlorine gas, calcium hypochlorite, or sodium hypochlorite. A process has been developed by Inco that involves the addition of SO_2 and air in the presence of a copper catalyst. Cyanide effluents can be acidified to allow volatilization and recovery of the cyanide followed by reneutralization of the effluent. Oxidation of cyanide by hydrogen peroxide has been used for the destruction of cyanide in gold mill effluents. Most of the chemical techniques have a number of problems: (1) complex metal cyanides are not completely degraded; (2) residual cyanide levels can still be too high (>0.5 ppm); and (3) other toxic reaction products can be formed.

A few species of fungi and bacteria are able to tolerate cyanide. The basis for resistance varies from the formation of a cyanide resistant cytochrome oxidase[68,69] to the utilization of cyanide as a carbon or nitrogen source.[70] Various fungi have been shown to metabolize cyanide by incorporating it into amino acids.[71,72] Fry and Millar[73] isolated the enzyme formamide hydrolase from the fungus *Stemphylium loti*, which converts cyanide to formamide. Powell et al.[74] isolated a strain of *Fusarium moniliforme* that degrades cyanide to ammonia and carbon dioxide via formamide and formate. *Escherichia coli*, *Chromobacterium violaceum*, *Bacillus* spp., and *Pseudomonas* spp. have been shown to degrade cyanide.[75-79]

Cyanide destruction by actinomycetes in trickling filters has been studied using both simple cyanides and metal cyanide complexes.[80] In filters acclimatized to cyanide, more than 99% of the cyanide was removed in solutions containing at least 100 mg L^{-1} of either free cyanide or as the cyanide complexes of zinc and cadmium.[80] Cyanide degradation capabilities of activated sludge digestors are similar to trickling filters.[81,82] In a laboratory study of an extended aeration modification of the activated sludge process, Gaudy et al.[83] found feeding the reactor with cyanide and glucose resulted in both cyanide degradation and nitrification. However each incremental increase in the cyanide concentration resulted in severe disturbances in the process performance. A mixture of coke plant wastewater and blast furnace blowdown water containing cyanide and phenols was treated using a fluidized bed reactor.[84] Petrochemical plants, steel manufacturing, and synthetic fuel processing generate wastewaters with high concentrations of phenols, cyanide, and ammonia each of which to some degree inhibits the degradation of the other components.[85] Richards and Shieh[86] used an anoxic-oxic activated sludge system with a high recycle rate. In the oxic phase, cyanides and thiocyanate are converted to ammonium, which is subsequently converted to nitrate. The nitrate is reduced to dinitrogen in the anoxic reactor with the phenols serving as carbon sources.

An innovative approach for cyanide removal involves the use of immobilized fungi such as *Stemphylium loti*, which contain the cyanide degrading enzyme formamide hydrolase in column reactors.[87] The advantages of the immobilized fungi are: (1) capital cost is low; (2) initial concentrations of up to 8000 ppm can be detoxified; and (3) no toxic reaction products are formed. The disadvantages of this system are: (1) the immobilized cells have limited stability; and (2) the enzyme will not hydrolyze metal cyanide complexes.

The Homestake Mine in Lead, South Dakota, is currently treating up to 21,000 m^3 of cyanide-containing effluent per day.[88] The wastewater is a mixture of underground mine water and decanted water from a tailings impoundment. A series of RBCs function in the removal of metals and cyanide destruction, nitrification, and denitrification. The system removed free cyanide, complexed cyanides, and thiocyanate by biodegradation, and metals were removed by biosorption.

3.5 Ammonium and Nitrate

Sequential ammonia oxidation and nitrate reduction reactions have been used to remove nitrogen from both municipal and industrial wastewaters. Ammonium nitrogen is oxidized to nitrate aerobically by nitrifying bacteria, then anaerobically reduced to nitrogen gas by the denitrifying bacteria, and thus removed from the effluent.

Ammonium is oxidized to nitrate by both chemolithotrophic bacteria, which require only inorganic compounds for growth, and heterotrophic bacteria, which require organic compounds for growth.[89] The chemolithotrophic bacteria are responsible for most of the nitrifying activity observed in nature. The chemolithotroph *Nitrosomonas* sp. oxidizes ammonium to nitrite, and *Nitrobacter* sp. oxidizes nitrite to nitrate.

$$NH_4^+ \rightarrow NO_2^- \rightarrow NO_3^-$$

The chemolithotrophic nitrifying bacteria can obtain all of their carbon by fixation of CO_2. *Nitrosomonas* has been shown to be inhibited by low concentrations of various organic compounds.[90] However, others[91,92] have shown that *Nitrosomonas* sp. can assimilate trace amounts of various organic compounds and are not inhibited by them. The nitrifiers also have very slow growth rates.[93,94]

Denitrification, which is also referred to as nitrate respiration, is mediated by a number of bacterial species.[95] *Pseudomonas*, *Paracoccus*, *Flavobacterium*, *Alcaligenes*, and *Bacillus* spp. are among the genera known to denitrify. The nitrate ion is reduced to dinitrogen gas by the following pathway:[95]

$$2NO_3^- \rightarrow 2NO_2^- \rightarrow 2NO \rightarrow N_2O \rightarrow N_2$$

Some species of bacteria are only capable of reducing one of the nitrogen oxides by one or two steps in the pathway to another intermediate. Actively denitrifying cultures are frequently a mixture in which the overall denitrifying activity is the result of several species each of which mediates one or more steps in the reaction sequence. Although the denitrifying bacteria are aerobic microorganisms, they can utilize oxidized nitrogen compounds as terminal electron acceptors in place of oxygen. Either low oxygen concentrations or the complete absence of oxygen is required for denitrification to occur.[96,97] Denitrifiers are generally nutritionally very versatile and can utilize a variety of carbon substrates.[98] Both nitrification and denitrification are optimal at pH values near neutrality.

Denitrification systems have been designed for industrial and municipal wastewater. Nitrogen concentrations in domestic wastewater are generally lower than those in industrial effluents. Two possible sequences for a nitrification/denitrification plant can be used.[99] The first, nitrification followed by denitrification usually requires the addition of an external carbon source to act as an electron donor for denitrification.[100] Methanol is the most frequently chosen carbon source due to its low cost and the low biomass yield from growth on this substrate. The second approach, denitrification followed by nitrification in which a large portion of the nitrified effluent is recycled back to the first reactor has several advantages: (1) organics in the effluent can be utilized as a carbon source for denitrification, which would reduce reagent costs, and (2) organics that might be toxic to the nitrifiers are removed in the first stage. The denitrification-nitrification sequence has the disadvantage of producing a final effluent containing appreciable nitrate concentrations and may not even be feasible if the waste stream is low in metabolizable organic compounds.

The types of reactors that can be used for nitrification and denitrification are the fluidized bed,[101,102] the CSTR,[103,104] the packed tower,[105] suspended sludge,[106] and the RBC.[107] The most popular systems for large scale wastewater nitrification are the CSTR, packed towers, and the RBC.[105] The CSTR provides the lowest ammonium concentration in the effluent, but is somewhat more difficult and expensive to operate than other systems. The packed tower and RBC produce higher ammonium concentrations in the effluent (1 to 3 mg N/L), but are relatively simple to control and operate. Both CSTRs and packed towers are used for denitrification, but the packed tower is preferable due to its simplicity of operation.

Two innovative systems that only require one reactor have been developed for the removal of nitrogen from wastewater. The LINPOR™ process uses an activated sludge reactor containing open pore plastic foam cubes. Nitrification occurs on the surface of the cubes, and denitrification occurs within the pores in the interior of the cubes.[108] Rogalla and Bourbigot[109] developed an upflow biofilter system that utilized a granular support system with aeration introduced at a point $1/3$ of the height of the column. The upper zone of the reactor was oxic, and the lower zone was anoxic so that both nitrification and denitrification could be accomplished in only one reactor. Another alternative to the activated sludge configuration is to immobilize bacteria such as the nitrifiers in carrageenan.[110] The immobilized bacteria will be more stable than activated sludge flocs. The recent advances occurring in reactor design and the microbiology of nitrification and denitrification will decrease the unit costs for the treatment of nitrogenous effluents.

Another application of the engineered ecosystem concept is the use of reed bed systems for the treatment of high strength nitrogenous wastewaters.[111] The root zone of the reeds has higher microbial activity than the surrounding matrix and also receives oxygen from the plant. Due to the high microbial activity and variations in oxygen concentration, organic matter decomposition, nitrification, and denitrification all occur in the root zone of the reeds.

4. GROUNDWATER

The groundwater in most areas of the world is contaminated to some degree with both organic and inorganic pollutants.[112] Polluted groundwater can be treated by pumping it to the surface for treatment and subsequently discharging it either into the subsurface again or into surface waters. Treatment can also be accomplished by stimulating microorganisms *in situ* to degrade the contaminants. *In situ* treatment is most suitable for organic compounds because most inorganic compounds must either be immobilized in place or removed from the groundwater. The most feasible technique for the removal of inorganic contaminants is to pump the groundwater to the surface for treatment. Among the inorganic pollutants found in groundwater are metals, radionuclides, selenium compounds, and inorganic nitrogen compounds.

Metals and radionuclides can potentially be removed from contaminated groundwater by biosorption. However, very little work has been done to determine the feasibility of using biosorbants for the removal of metals from groundwater. Groundwater recharge as a means of diluting contaminants in groundwater and replenishing depleted aquifers is currently being used. Municipal wastewater in Riverside County, California, is used for aquifer recharge once it has been treated in a fluidized biofilm reactor in order to remove inorganic nitrogen by nitrification-denitrification.[113]

Leachate from landfills is a major source of both organic and inorganic pollutants in groundwater. The composition of landfill leachate depends on the type of solid waste, the environmental conditions of the fill, age of the fill, characteristics of the incoming water, and the type of soil.[114] Leachates from younger landfills are high in biodegradable carbon concentrations, but low in ammonium ion concentrations. The converse is true for leachates from older landfills. Elefsiniotis et al.[114] used two laboratory scale CSTRs utilizing an anoxic reactor followed in sequence by an oxic reactor to remove nitrogen from landfill leachate. Because the leachate was from an older fill and was low in nutrients, it was necessary to add phosphate and methanol to the reactor. The most important process variable was determined to be the recycle ratio with a ratio of 6:1 to be optimum. Mennerich and Albers[115] compared two systems for the treatment of a leachate from a young landfill. The leachate was subjected to an anaerobic pretreatment step. A single stage activated sludge system with a lower anoxic zone was compared to an RBC with a preceding denitrifying step. For that particular leachate, the activated sludge system was found to be satisfactory, whereas the RBC was not.

5. SOLID WASTES

Metals are the inorganic contaminants most frequently found in soils and solid wastes. Most metal-contaminated soils are currently being landfilled, but more stringent environmental regulations will require that alternate methods be developed.[116] Immobilization of metals in contaminated soils using cement or other fixing agents has been investigated.[117] The other alternative is the removal of metals from contaminated soils or solid wastes to produce a concentrated waste. The metals in the resulting waste can either be disposed of or reused. Volatilization of metals from soil by low temperature incineration in a rotary kiln incinerator has been considered.[118] Various leaching techniques have been evaluated for the removal of heavy metals from soil. Hydrochloric acid and nitrilotriacetic acid are among the leaching agents used.[119,120] Once metals have been leached, they could be concentrated by conventional means or by biosorption.

One innovative biotechnological approach to the removal of metals from contaminated solids is to mix the waste with sulfur and inoculate with *Thiobacillus ferrooxidans T. thiooxidans* oxidizes sulfur to sulfuric acid, which thus acts as an acid leaching agent. Jack et al.[121] added sulfur and a bacterial inoculum to petroleum cokes rich in trace metals in order to recover vanadium. The major difficulty encountered was toxicity to *T. thiooxidans* as a result of the chromate and vanadate ions leached out of the coke. Olson et al.[122] used *Thiobacillus ferrooxidans* in a mineral medium containing pyrite as an energy source to generate a lixiviant to leach cobalt from smelter wastes. Bacterially generated acid solutions have been used to leach copper from glass manufacturing wastes[123] and various other metals from nonsulfide waste products such as slag, galvanic sludge, and fly ash.[124] However, it is not possible to bacterially remove metals such as lead, which form insoluble sulfates. To date, all of these studies have been laboratory scale experiments, and no industrial or pilot scale facilities have been constructed to exploit this technology.

Bacterial leaching using *T. ferrooxidans* has been used on a large scale for the recovery of copper and uranium from low grade sulfide ores.[125–128] Economically valuable metals such as copper and nickel often occur as sulfides mixed with iron sulfides such as pyrite (FeS_2). Metals are solubilized both by direct bacterial oxidation and also by the ferric ion, which is also bacterially generated. The following bacterial and chemical reactions are involved in the bacterial leaching of metals.

Chemical:

1. $MS + 2Fe^{3+} \rightarrow M^{2+} + 2Fe^{2+} + S^{\circ}$

Bacterial:

1. $S^{\circ} + 3/2\ O_2 + H_2O \rightarrow H_2SO_4$
2. $MS + 2O_2 \rightarrow MSO_4$
3. $4Fe^{2+} + 4H^+ + O_2 \rightarrow 4Fe^{3+} + 2H_2O$

Techniques for contacting ores with the leaching solution include dump leaching, heap leaching, and tank leaching.[125] Leach dumps are usually located in valleys in which low grade ores of varying particle size (<1.0 m diameter) are dumped in lifts of 50 to 100 ft in height. Copper leach dumps can contain up to 4 billion tons of material. Leach solutions are introduced on the surface of the dump by sprinkling or flooding. The solution is recovered from the base of the dump and processed to recover the metals. In a heap leaching operation, the ore is placed on an impermeable surface, and the leaching solution is applied to the top of the heap. Approximately 100,000 to 500,000 tons of ore are used in a single heap. Tank leaching can be done in a standard bioreactor such as a CSTR. The technology for the bacterial leaching of metals is well developed and could be a technically and economically feasible method for the removal of metals from contaminated sludges and soils.

6. CONCLUSIONS AND FUTURE POSSIBILITIES

Biotechnological methods for the treatment of inorganic wastes are at different stages of development for each type of waste. Bioreactors of various designs have been used in the fermentation industry and also for the treatment of organic wastes. Several of the commonly used reactor designs could be used to contact liquid effluents containing inorganic pollutants with biomass in order to treat the effluents.

Considering the inherent advantages in specificity and binding efficiency, biosorbents have excellent potential for the removal of metals and radionuclides from wastewater, particularly for effluents with low concentrations of metals. The preferred technique for biosorption is to use the ion exchange configuration. A large number of laboratory scale studies and some pilot plant studies have been carried out to evaluate biosorption, but no commercial scale treatment plants utilizing biosorption are presently in use. The major problems to be overcome in the development of biosorbent-based processes are cost, loading capacity, and physical and chemical stability.

Although the use of engineered ecosystems for the treatment of acid mine drainage has only begun to be tested at the pilot stage, it has the potential to be an economically viable long-term solution to the problem of acid mine drainage. Acid mine drainage can be ameliorated by natural processes that occur in wetlands such as sulfate reduction to produce hydrogen sulfide, which precipitates the metals in the drainage. Because the biological treatment of acid mine drainage is not always successful and the mechanisms of amelioration are poorly understood, more basic research is necessary in this field.

The bacterial removal of selenium from wastewaters and groundwater has only been tested at the pilot plant stage. The biochemical mechanisms for biological selenium reduction are poorly understood. If the mechanisms of microbial selenium reduction could be elucidated, it may be possible to design an analogous chemical system for selenium removal that would be more efficient than the processes currently in use. Future work should be directed towards basic research on selenium biochemistry and testing larger scale biological processes for selenium removal.

Large volumes of wastewater containing low concentrations of cyanide can be treated using RBCs. CSTRs have been used for the degradation of wastes containing cyanide and phenols. Future research should be directed towards the development of processes for the treatment of effluents containing higher cyanide concentrations and metal cyanide complexes.

The most satisfactory method for the removal of inorganic nitrogenous compounds from wastewater is the combination of nitrification and denitrification. There are a number of reactor configurations used for the large scale removal of nitrogen from wastewater. The main challenge in nitrogen removal is to improve existing technology in order to reduce the unit costs of the technology currently used.

At the present time, there are few applications for the use of biotechnology in the treatment of solid inorganic wastes. However considerable potential exists in using bacterially generated sulfuric acid to leach toxic metals from contaminated soils and sludges.

7. ACKNOWLEDGMENTS

The authors gratefully acknowledge the editorial assistance of D. Limoges and graphic arts assistance of Jean Benedict.

8. REFERENCES

1. Cerjan-Stefanovic, S., F. Briski, and M. Kastelan-Macan. "Separation of Silver from Waste Waters by Ion-Exchange Resins and Concentration by Microbial Cells. Silver Uptake by Microbial Cells from Treated Waste Waters After Ion Exchange," *Fresenius J. Anal. Chem.* 339:636–639 (1991).

2. McCready, R. G. L., J. Salley, and W. D. Gould. "Biorecovery of Selenium from Smelter Effluents," in *Proceedings Pacific Rim Congress 90. Gold Coast, Queensland, Australia, 6–12 May 1990, Volume II* (Parkville, Australia: The Australasian Institute of Mining and Metallurgy, 1990), pp. 623–626."

3. Hynes, T. P., R. M. Schmidt, T. Meadley, and N. A. Thompson. "The Impact of Effluents from a Uranium Mine and Mill Complex in Northern Saskatchewan on Contaminant Concentrations in Receiving Waters and Sediments," *Water Pollut. Res. J. Canada* 22:559–569 (1987).

4. Beveridge, T. J., and R. G. E. Murray. "Uptake and Retention of Metals by Cell Walls of *Bacillus subtilis*," *J. Bacteriol.* 127:1502–1518 (1976).

5. Beveridge, T. J., and R. G. E. Murray. "Sites of Metal Deposition in the Cell Wall of *Bacillus subtilis*," *J. Bacteriol.* 141:876–887 (1980).

6. Tsezos, M., R. G. L. McCready, and J. P. Bell. "The Continuous Recovery of Uranium from Biologically Leached Solutions Using Immobilized Biomass," *Biotechnol. Bioeng.* 34:10–17 (1989).

7. Venkobachar, C. "Metal Removal by Waste Biomass to Upgrade Wastewater Treatment Plants," *Water Sci. Technol.* 22:319–320 (1990).

8. Byerley, J. J., J. M. Scharer, and A. M. Charles. "Uranium (VI) Biosorption from Process Solutions," *The Chem. Eng. J.* 36:B49–B57 (1987).

9. Huang, C.-P., C.-P. Huang, and A. L. Morehart. "The Removal of Cu (II) from Dilute Aqueous Solutions by *Saccharomyces cerevisiae*," *Water Res.* 24:433–439 (1990).

10. Friis, N., and P. Myers-Keith. "Biosorption of Uranium and Lead by *Streptomyces longwoodensis*," *Biotechnol. Bioeng.* 28:21–28 (1986).

11. Aksu, Z., and T. Kutsal. "A Comparative Study for Biosorption Characteristics of Heavy Metal Ions with *C. vulgaris*," *Environ. Technol.* 11:979–987 (1990).

12. Greene, B., M. T. Henzl, J. M. Hosea, and D. W. Darnall. "Elimination of Bicarbonate Interference in the Binding of U (VI) in Mill-Waters to Freeze Dried *Chlorella vulgaris*," *Biotechnol. Bioeng.* 28:764–767 (1986).

13. Harris, P. O., and G. J. Ramelow. "Binding of Metal Ions by Particulate Biomass Derived from *Chlorella vulgaris* and *Scenedesmus quadricauda*," *Environ. Sci. Technol.* 24:220–228 (1990).

14. Scott, C. D. "Removal of Dissolved Metals by Plant Tissue," *Biotechnol. Bioeng.* 39:1064–1068 (1992).

15. Kasan, H. C., and A. A. Baecker. "An Assessment of Toxic Metal Biosorption by Activated Sludge from the Treatment of Coal-Gasification Effluent of a Petrochemical Plant," *Water Res.* 23:795–800 (1989).

16. Hunt, S. "Diversity of Biopolymer Structure and its Potential for Ion Binding Applications," in *Immobilization of Ions by Bio-Sorption*, H. Eccles and S. Hunt, Eds. (Chichester, U.K.: Society of Chemical Industry, Ellis Horwood Ltd., 1986), pp. 15–46.

17. Jang, L. K., S. L. Lopez, S. L. Eastman, and P. Pryfogle. "Biorecovery of Copper and Cobalt by Biopolymer Gels," *Biotechnol. Bioeng.* 37:266–273 (1991).

18. Neilands, J. B. "Microbial Iron Compounds," *Ann. Rev. Biochem.* 50:715–731 (1981).

19. Awadalla, F. T., and B. Pesic. "Biosorption of Cobalt with the AMT™ Metal Removing Agent," *Hydrometallurgy* 28:65–80 (1992).

20. Tsezos, M. "The Role of Chitin in Uranium Adsorption by *R. arrhizus*," *Biotechnol. Bioeng.* 25:2025–2040 (1983).

21. Tsezos, M., and B. Volesky. "The Mechanism of Uranium Biosorption by *Rhizopus arrhizus*," *Biotechnol. Bioeng.* 24:385–401 (1982).

22. Tsezos, M., and S. Mattar. "A Further Insight into the Mechanism of Biosorption of Metals by Examining Chitin EPR Spectra," *Talanta* 33:225–232 (1986).

23. Glombitza, F., and U. Iske. "Treatment of Biomass for Increasing Biosorption Activity," in *Biohydrometallurgy: Proceedings of the International Symposium, Jackson Hole, Wyoming, August 13–18, 1989, CANMET SP89–10*, J. Salley, R. G. L. McCready, and P. L. Wichlacz, Eds. (Ottawa: Canada Centre for Mineral and Energy Technology, 1989), pp. 329–340.

24. Pighi, L., T. Pumpel, and F. Schinner. "Selective Accumulation of Silver by Fungi," *Biotechnol. Lett.* 11:275–280 (1989).

25. Muzzarelli, R. A. A., F. Bregani, and F. Sigon. "Chelating Capacities of Amino Acid Glucans and Sugar Acid Glucans Derived from Chitosan," in *Immobilization of Ions by Bio-Sorption*, H. Eccles and S. Hunt, Eds. (Chichester, U.K.: Society of Chemical Industry, Ellis Horwood Ltd., 1986), pp. 173–182.

26. Costerton, J. W. "The Development of Bacterial Biofilms for the Selective Removal of Metals from Mining Effluents," DSS Contract No. 23440-9-9304 (1991).

27. Morper, M. "Anaerobic Sludge — A Powerful and Low-Cost Sorbent for Heavy Metals," in *Immobilization of Ions by Bio-Sorption*, H. Eccles and S. Hunt, Eds. (Chichester, U.K.: Society of Chemical Industry, Ellis Horwood Ltd., 1986), pp. 91–99.

28. Tsezos, M., S. H. Noh, and M. H. I. Baird. "A Batch Reactor Mass Transfer Kinetic Model for Immobilized Biomass Biosorption," *Biotechnol. Bioeng.* 32:545–553 (1988).

29. Klein, J., and K. D. Vorlop. "Immobilization Techniques — Cells," in *Comprehensive Biotechnology, Vol. 2*, M. Moo-Young, Ed. (New York: Pergamon Press, 1985), pp. 203–224.

30. Kiff, R. J., and D. R. Little. "Biosorption of Heavy Metals by Immobilized Fungal Biomass," in *Immobilization of Ions by Bio-Sorption*, H. Eccles and S. Hunt, Eds. (Chichester, U.K.: Society of Chemical Industry, Ellis Horwood Ltd., 1986), pp. 71–80.

31. Jeffers, T. H., C. R. Ferguson, and D. C. Seidel. "Biosorption of Metal Contaminants Using Immobilized Biomass," in *Biohydrometallurgy: Proceedings of the International Symposium, Jackson Hole, Wyoming, August 13–18, 1989, CANMET SP89–10*, J. Salley, R. G. L. McCready, and P. L. Wichlacz, Eds. (Ottawa: Canada Centre for Mineral and Energy Technology, 1989), pp. 317–327.

32. McEldowney, S. "Microbial Biosorption of Radionuclides in Liquid Effluent Treatment," *Appl. Biochem. Biotechnol.* 26:159–179 (1990).

33. Tsezos, M., and D. M. Keller. "Adsorption of Radium-226 by Biological Origin Absorbents," *Biotechnol. Bioeng.* 25:201–215 (1983).

34. Tsezos, M., M. H. I. Baird, and L. W. Shemilt. "The Kinetics of Radium Biosorption," *Chem. Eng. J.* 33:B35–B41 (1986).

35. Tsezos, M., M. H. I. Baird, and L. W. Shemilt. "The Use of Immobilized Biomass to Remove and Recover Radium from Elliot Lake Uranium Tailing Streams," *Hydrometallurgy* 17:357–368 (1987).

36. Kalin, M., and W. N. Wheeler. "Algal Biopolishing of Zinc," *Final Report to CANMET*, Department of Energy Mines and Resources Canada, DSS Contract No. 23440-1-9009/01SQ (1992).

37. Hedin, R. S., D. M. Hyman, and R. W. Hammack. "Implications of Sulfate-Reduction and Pyrite Formation Processes for Water Quality in a Constructed Wetland: Preliminary Observations," in *Mine Drainage and Surface Mine Reclamation, Vol. I*, Bureau of Mines I.C. 9183, U.S. Department of the Interior (1988), pp. 382–388.

38. Tuttle, J. H., P. R. Dugan, and C. I. Randles. "Microbial Sulfate Reduction and Its Potential Utility as an Acid Mine Water Pollution Abatement Procedure," *Appl. Microbiol.* 17:297–302 (1969a).

39. Tuttle, J. H., P. R. Dugan, C. B. MacMillan, and C. I. Randles. "Microbial Dissimilatory Sulfur Cycle in Acid Mine Water," *J. Bacteriol.* 97: 594–602 (1969b).

40. Wakao, N., T. Takashi, Y. Sakurai, and H. Shiota. "A Treatment of Acid Mine Water Using Sulfate-Reducing Bacteria," *J. Ferment. Technol.* 57:445–452 (1979).

41. Gale, N. L., B. G. Wixson, M. G. Hardie, and J. C. Jennet. "Aquatic Organisms and Heavy Metals in Missouri's New Lead Belt," *Water Resour. Bull. Amer. Water Resour. Assoc.* 9:673–688 (1973).

42. Gale, N. L., and B. G. Wixson. "Removal of Heavy Metals from Industrial Effluents by Algae," *Dev. Indust. Microbiol.* 20:259–274 (1978).

43. P. Lane and Associates Ltd. "Heavy Metal Removal from Gold Mining and Tailings Effluents Using Indigenous Aquatic Macrophytes," *Phase 1. Report to CANMET*, Department of Energy Mines and Resources, DSS Contract No. 23440-7-9173/ 015Q (1989).

44. Murray, E. W., S. P. Goudey, R. G. L. McCready, and J. Salley. "Laboratory and Field Testing of Salt-Supplemented Clay Cap as an Impermeable Seal Over Pyritic Slates," in *Mine Drainage and Surface Mine Reclamation Proceedings of a Conference Pittsburgh, Pennsylvania on April 19–21, 1988, Vol. 1*, U.S. Bureau of Mines Information Circular 9183 (1988), pp. 52–58.

45. Doran, J. W. "Microorganisms and the Biological Cycling of Selenium," in *Advances in Microbial Ecology, Vol. 6*, K. L. Marshall, Ed. (New York: Plenum Press, Inc., 1982), pp. 1–32.

46. Stadtman, T. C. "Selenium Biochemistry," *Ann. Rev. Biochem.* 59:111–127 (1990).

47. Hoffmann, J. E. "Selenium and Tellurium — Rare but Ubiquitous," *J. Metals* 41:32 (1989).

48. U.S. Department of the Interior. "Selenium and Tellurium 1990," Mineral Industry Surveys, U.S. Bureau of Mines (1991).

49. Altringer, P. B., D. M. Larsen, and K. R. Gardiner. "Bench Scale Process Development of Selenium Removal from Wastewater Using Facultative Bacteria," in *Biohydrometallurgy: Proceedings of the International Symposium, Jackson Hole, Wyoming, August 13–18, 1989, CANMET SP89–10*, J. Salley, R. G. L. McCready, and P. L. Wichlacz, Eds. (Ottawa: Canada Centre for Mineral and Energy Technology, 1989), pp. 643–657.

50. Primak, J., and J. Pageau. "Industrial Waste Treatment Plant Performance," in *Environmental Controls in Metallurgical Industries and Scrap Metal Recycling*, M. E. Chalkley and A. J. Oliver, Eds. (Ottawa: Conference of Metallurgists of CIM, 1991), pp. 219–230.

51. Marchant, W. N., R. O. Dannenberg, and P. T. Brooks. "Selenium Removal from Acidic Waste Water Using Zinc Reduction and Lime Neutralization," U.S. Bureau of Mines, Report of Investigations RI-8312 (1978).

52. Merrill, D. T., M. A. Manzone, J. J. Petersen, and D. S. Parker. "Field Evaluation of Arsenic and Selenium Removal by Iron Coprecipitation," prepared for Electric Power Research Institute by Brown and Caldwell, Research Report 910–3 (1987).

53. Wilmoth, B. C., T. L. Baugh, and D. W. Decker. "Removal of Selected Trace Elements from Acid Mine Drainage Using Existing Technology," proceedings of the 33rd Purdue Industrial Waste Conference, School of Civil Engineering, Purdue University, Ann Arbor, MI (1978), pp. 886–894.

54. Trussell, R. R., A. Trussell, P. Kreft, and J. M. Montgomery. "Selenium Removal from Ground Water Using Activated Alumina," U.S. Environmental Protection Agency, Report EPA-600-12-80-153 (1980).

55. Ahlgren, R. M. "Membrane vs. Resinous Ion Exchange Demineralization," *Ind. Water Eng.* 8:12–14 (1969).

56. Barkes, L., and R. W. Fleming. "Production of Dimethyl Selenide Gas from Inorganic Selenium by Eleven Soil Fungi," *Bull. Environ. Contam. Toxicol.* 12:308–311 (1974).

57. Doran, J. W., and M. Alexander. "Microbial Transformations of Selenium," *Appl. Environ. Microbiol.* 33:31–37 (1977).

58. Maiers, D. T., P. L. Wichlacz, D. L. Thompson, and D. F. Bruhn. "Selenate Reduction by Bacteria from a Selenium Rich Environment," *Appl. Environ. Microbiol.* 54:2591–2593 (1988).

59. McCready, R. G. L., J. N. Campbell, and J. I. Payne. "Selenite Reduction by *Salmonella heidelberg*," *Can. J. Microbiol.* 12:703–714 (1966).

60. Gerrard, T. L., J. N. Telford, and H. H. Williams. "Detection of Selenium Deposits in *Escherichia coli* by Electron Microscopy," *J. Bacteriol.* 119:1057–1060 (1974).

61. Silverberg, B. A., P. T. S. Wong, and Y. K. Chau. "Localization of Selenium in Bacterial Cells Using TEM and Energy Dispersive X-ray Analysis," *Arch. Microbiol.* 107:1–6 (1976).

62. Falcone, G., and W. J. Nickerson. "Reduction of Selenite by Intact Yeast Cells and Cell Free Preparations," *J. Bacteriol.* 85:763–771 (1963).

63. Macy, J. M., T. A. Michel, and D. G. Kirsch. "Selenate Reduction by a *Pseudomonas* Species: A New Mode of Anaerobic Respiration," *FEMS Microbiol. Lett.* 61:195–198 (1989).

64. Steinberg, N. A., and R. S. Oremland. "Dissimilatory Selenate Reduction Potentials in a Diversity of Sediment Types," *Appl. Environ. Microbiol.* 56:3550–3557 (1990).

65. Oremland, R. S., J. T. Hollibaugh, A. S. Maest, T. S. Presser, L. G. Miller, and C. W. Culbertson. "Selenate Reduction to Elemental Selenium by Anaerobic Bacteria in Sediments and Culture: Biogeochemical Significance of a Novel, Sulfate-Independant Respiration," *Appl. Environ. Microbiol.* 55: 2333–2343 (1989).

66. Larsen, D. M., K. R. Gardiner, and P. B. Altringer. "Biologically Assisted Control of Selenium in Process Waste Waters" in *Biotechnology in Minerals and Metal Processing*, B. Scheiner, F. Doyle, and S. K. Kawatra, Eds. (Littleton, CO: Society of Mining Engineers, Inc., 1989), pp. 177–186.

67. Gerhard, M. B., F. B. Green, R. D. Newman, T. J. Lundquist, R. B. Tresan, and W. J. Oswald. "Removal of Selenium Using a Novel Algal-Bacterial Process," *Res. J. WPCF* 63:799–805 (1991).

68. Arima, K., and R. Oka. "Cyanide Resistance in Achromobacter. I. Induced Formation of Cytochrome a2 and Its Role in Cyanide Resistant Respiration," *J. Bacteriol.* 90:734–743 (1965).

69. Oka, T., and K. Arima. "Cyanide Resistance in Achromobacter. II. Mechanism of Cyanide Resistance," *J. Bacteriol.* 90:744–747 (1965).

70. Knowles, C. "Microorganisms and Cyanide," *Bacteriol. Rev.* 40:652–700 (1976).

71. Allen, J., and G. A. Strobel. "The Assimilation of HCN by a Variety of Fungi," *Can. J. Microbiol.* 12:414–416 (1966).

72. Strobel, G. A. "4-Amino-4-Cyanobutyric Acid as an Intermediate in Glutamate Biosynthesis," *J. Biol. Chem.* 242:3265–3269 (1967).

73. Fry, W. E., and Millar, R. L. "Cyanide Degradation by an Enzyme from *Stemphylium loti*," *Arch. Biochem. Biophys.* 151:468–474 (1972).

74. Powell, K. A., A. J. Beardsmore, T. W. Naylor, and E. G. Corcoran. "The Microbial Treatment of Cyanide Waste," in *Effluent Treatment in the Process Industries*, European Federation of Chemical Engineering, Publication Series No. 31, Event No. 282 of the EFCE, I. Chem. Eng. Symposium Series, No.77, (New York: Pergamon Press, Inc., 1983), pp. 305–313.

75. Brysk, M. M., W. A. Corpe, and L. V. Hankes. "β-Cyanoalanine Formation in *Chromobacterium violaceum*," *J. Bacteriol.* 97:322–327 (1969).

76. Skowronski, B., and G. A. Strobel. "Cyanide Resistance and Cyanide Utilization by a Strain of *Bacillus pumilis*," *Can. J. Microbiol.* 15:93–98 (1969).

77. Dunnill, P. M., and L. Fowden. "Enzymic Formation of Cyanoalanine from Cyanide by *Escherichia coli* Extracts," *Nature* 208:1206–1208 (1965).

78. Castric, P. A., and Strobel, G. A. "Cyanide Metabolism by *Bacillus megaterium*," *J. Biol. Chem.* 244:4089–4090 (1969).

79. Mudder, T. I., and J. L. Whitlock. "Strain of *Pseudomonas paucimobilis*," U.S. Patent No. 4,461,834, July 24, 1984.

80. Pettet, A., and E. V. Mills. "Biological Treatment of Cyanides with and without Sewage," *J. Appl. Chem.* 4:434–443 (1954).

81. Fuji, Y., and T. Oshimi. "Process for Treating Wastewater Containing Nitriles," U.S. Patent No. 3,756,947, September 4, 1973.

82. Kato, A. H., and K. K. Yamamura. "Treating Waste Water Containing Nitriles and Cyanides," U.S. Patent No. 3,940,332, February 24, 1976.

83. Gaudy, A. F., Jr., E. T. Gaudy, Y. J. Feng, and G. Brueggemann. "Treatment of Cyanide Waste by the Extended Aeration Process," *J. Water Pollut. Control Fed.* 54:153–164 (1983).

84. Nutt, S. G., H. Melcer, and J. H. Pries. "Two-Stage Biological Fluidized Bed Treatment of Coke Plant Wastewater for Nitrogen Control," *J. Water Pollut. Control Fed.* 56:851–857 (1984).

85. Shivaraman, N., and N. M. Parhad. "Biodegradation of Cyanide by *Pseudomonas acidovorans* and Influence of pH and Phenol," *Indian J. Environ. Health* 27:1–8 (1985).

86. Richards, D. J., and W. K. Shieh. "Anoxic-Oxic Activated-Sludge Treatment of Cyanides and Phenols," *Biotechnol. Bioeng.* 33:32–38 (1989).

87. Nazly, N., C. J. Knowles, A. J. Beardsmore, W. T. Naylor, and E. G. Corcoran. "Detoxification of Cyanide by Immobilized Fungi," *J. Chem. Tech. Biotechnol.* 33B:119–126 (1983).

88. Whitlock, J. L., and G. R. Smith. "Operation of Homestake's Cyanide Biodegradation Wastewater System Based on Multi-Variable Trend Analysis," in *Biohydrometallurgy: Proceedings of the International Symposium, Jackson Hole, Wyoming, August 13–18, 1989, CANMET SP89–10*, J. Salley, R. G. L. McCready, and P. L. Wichlacz, Eds. (Ottawa: Canada Centre for Mineral and Energy Technology, 1989), pp. 613–625.

89. Schmidt, E. L. "Nitrification in Soil," in *Nitrogen in Agricultural Soils*, F. J. Stevenson, Ed. (Madison, WI: American Society of Agronomy, Inc., 1982), pp. 253–288.

90. Winogradsky, S. N. "Microbiologie du Sol: Problemes et Methodes," (Paris: Masson et Cie., 1949).

91. Clark, C., and E. L. Schmidt. "Effect of Mixed Culture on *Nitrosomonas europeae* Stimulated by Uptake and Utilization of Pyruvate," *J. Bacteriol.* 91:367–373 (1966).

92. Clark, C., and E. L. Schmidt. "Growth Response of *Nitrosomonas europeae* to Amino Acids," *J. Bacteriol.* 93:1302–1308 (1967).

93. Belser, L. W. "Nitrate Reduction to Nitrite, a Possible Source of Nitrite for Nitrite-Oxidizing Bacteria," *Appl. Environ. Microbiol.* 34:403–410 (1977).

94. Bock, E. "Lithotrophic and Chemoautotrophic Nitrifying Bacteria," in *Microbiology 1978*, D. Schlessinger, Ed. (Washington, DC: American Society for Microbiology, 1978), pp. 310–314.

95. Payne, W. J. *Denitrification.* (New York: John Wiley and Sons, Inc., 1981).

96. Nelson, L. M., and R. Knowles. "Effect of Oxygen and Nitrate on Nitrogen Fixation and Denitrification by *Azospirillum brasilense* Grown in Continuous Culture," *Can. J. Microbiol.* 24:1395–1403 (1978).

97. Terai, H., and T. Mori. "Studies on Phosphorylation Coupled with Denitrification and Aerobic Respiration in *Pseudomonas denitrificans*," *Bot. Mag.* 38:231–244 (1975).

98. Buchanan, R. E., and N. E. Gibbons, Eds. *Bergey's Manual of Determinative Bacteriology, 8th ed.* (Baltimore, MD: Williams and Wilkins, Inc., 1974).

99. Furun, L., E. S. K. Chian, and W. H. Gross. "Nitrogen Removal from Coal Gasification Waste Water by Biological Treatment," *Water Treat.* 4:483–493 (1989).

100. Narkis, N., M. Rebhun, and C. Sheindorf. "Denitrification at Various Carbon to Nitrogen Ratios," *Water Res.* 13:93–98 (1979).

101. Turner, C. D., C. S. Ong, and J. R. Gallagher. "Coupled Biological Downflow Fluid Bed Reactor Treatment of Synfuels Wastewater," in *Proceedings of the 43rd Purdue Industrial Waste Conference* (Chelsea, MI: Lewis Publishers Inc., 1989), pp. 469–477.

102. Melcer, H., and S. G. Nutt. "Nitrogen Control of Complex Industrial Wastewaters," *J. Environ. Eng.* 114:166–178 (1988).

103. Neytzell-DeWilde, F. G. "Treatment of Effluents from Ammonia Plants. I. Biological Nitrification of an Inorganic Effluent from a Nitrogen-Chemicals Complex," *Water SA* 3:113–122 (1977).

104. Neytzell-DeWilde, F. G., G. R. Nurse, and J. Groves. "Treatment of Effluents from Ammonia Plants. IV. Denitrification of an Inorganic Effluent from a Nitrogen-Chemicals Complex Using Methanol as Carbon Source," *Water SA* 3:142–154 (1977).

105. Grady, C. P. L., Jr., and H. C. Lim. *Biological Wastewater Treatment: Theory and Applications* (New York: Marcel Dekker, Inc., 1980).

106. Bridle, T. R., D. C. Climenhage, and A. Stelzig. "Operation of a Full Scale Nitrification-Denitrification Industrial Waste Treatment Plant," *J. Water Pollut. Control Fed.* 51:127–139 (1979).

107. Sforza, M. P., M. Catano, and F. Pierucci. "Study of Integrated Biodisc Process for Coke-Oven Wastewaters Treatment," *Water Supply* 6:175–180 (1987).

108. Reimann, H. "The LINPOR-Process for Nitrification and Denitrification," *Water Sci. Tech.* 22:297–298 (1990).

109. Rogalla, F., and M.-M. Bourbigot. "New Developments in Complete Nitrogen Removal with Biological Aerated Filters," *Water Sci. Tech.* 22:273–280 (1990).

110. Wijffels, R. H., C. D. de Gooijer, S. Kortekaas, and J. Tramper. "Growth and Substrate Consumption of *Nitrobacter agilis* Cells Immobilized in Carrageenan. II. Model Evaluation," *Biotechnol. Bioeng.* 38:232–240 (1991).

111. Job, G. D., A. J. Biddlestone, and K. R. Gray. "Treatment of High Strength Agricultural and Industrial Effluents Using Reed Bed Treatment Systems," *TransIChemE* 69:187–189 (1991).

112. Crawford, R. L. "Bioremediation of Groundwater Pollution," *Curr. Opinion Biotechnol.* 2:436–439 (1991).

113. MacDonald, D. V. "Denitrification by Fluidized Biofilm Reactor," *Water Sci. Technol.* 22:451–461 (1990).

114. Elefsiniotis, P., R. Manoharan, and D. S. Mavinic. "The Effects of Sludge Recycle Ratio on Nitrification-Denitrification Performance in Biological Treatment of Leachate," *Environ. Technol. Lett.* 10:1041–1050 (1989).

115. Mennerich, A., and H. Albers. "Nitrification/Denitrification of Landfill Leachates," *Water Supply* 6:157–166 (1988).

116. Esposito, P., J. Hessling, B. B. Locke, M. Taylor, M. Szabo, R. Thurnau, C. Rogers, R. Traver, and E. Barth. "Results of Evaluations of Contaminated Soil Treatment Methods in Conjunction with the CERCLA BDAT Program," proceedings of the 81st Annual Meeting of APCA Dallas, Texas, June 19–24 (1988), pp. 1–25.

117. Akhter, H., L. G. Butler, S. Branz, F. K. Cartledge, and M. E. Tittlebaum. "Immobilization of As, Cd, Cr and Pb-Containing Soils by Using Cement or Pozzolanic Fixing Agents," *J. Hazardous Materials* 24:145–155 (1990).

118. Lighty, J. S., E. G. Eddings, E. R. Lingren, D. Xiao-Xue, D. W. Pershing, R. M. Winter, and W. H. McClennen. "Rate Limiting Processes in the Rotary-Kiln Incineration of Contaminated Solids," *Combust. Sci. and Tech.* 74:31–49 (1990).

119. Tuin, B. J. W., and Tels, M. "Extraction Kinetics of Six Heavy Metals from Contaminated Clay Soils," *Environ. Technol.* 11:541–554 (1990).

120. Linn, J. H., and Elliott, H. A. "Mobilization of Cu and Zn in Contaminated Soil by Nitrilotriacetic Acid," *Water, Air, and Soil Pollut.* 37:449–458 (1988).

121. Jack T. R., E. A. Sullivan, and J. E. Zajic. "Growth Inhibition of *Thiobacillus thiooxidans* by Metals and Reductive Detoxification of Vanadium (V)," *European J. Appl. Microbiol. Biotechnol.* 9: 21–30 (1980).

122. Olson, G. J., C. K. Sakai, E. J. Parks, and F. E. Brinckman. "Bioleaching of Cobalt from Smelter Wastes by *Thiobacillus ferrooxidans,*" *J. Ind. Microbiol.* 6:49–52 (1990).

123. Clark, T. R., and H. L. Ehrlich. "Copper Removal from an Industrial Waste by Bioleaching," *J. Ind. Microbiol.* 9:213–218 (1992).

124. Bosecker, K. "Bacterical Metal Recovery and Detoxification of Industrial Waste," in *Workshop on Biotechnology for the Mining, Metal-Refining and Fossil Fuel Processing Industries, Biotechnology and Bioengineering Symposium No. 16,* H. L. Ehrlich and D. S. Holmes, Eds. (New York: John Wiley and Sons, 1986), pp. 105–120.

125. Brierley, C. L. "Bacterial Leaching," *CRC Crit. Rev. Microbiol.* 6:207–262 (1978).

126. Lundgren, D. G., and M. Silver. "Ore Leaching by Bacteria," *Ann. Rev. Microbiol.* 34:263–283 (1980).

127. McCready, R. G. L., and W. D. Gould. "Bioleaching of Uranium," in *Microbial Mineral Recovery,* H. L. Ehrlich and C. L. Brierley, Eds. (New York: McGraw-Hill, Inc., 1990), pp. 107–125.

128. Wadsworth, M.E. "Recovery of Metal Values from Low Grade Copper Sulfide Ores," *Sep. Sci. and Technol.* 16:1081–1112 (1981).

CHAPTER 5

Physiology of Biodegradative Microorganisms

Peter Adriaens and William J. Hickey

1. INTRODUCTION

Bioremediation processes capitalize on the activities of aerobic or anaerobic heterotrophic microorganisms. Microbial activity is affected both qualitatively and quantitatively by a number of physicochemical environmental parameters, often referred to as ecophysiological variables. The range of ecophysiological factors is quite broad and varies depending on the environment under consideration. An exhaustive review of ecophysiological factors will not be presented here; the interested reader is referred to several existing publications for comprehensive reviews.[1,2] The present discussion will instead focus on those factors having the most direct impact on bioremediation activities, which have some potential for manipulation or may serve as indicators of the occurrence or types of biological transformations. Hence, the roles of energy sources, electron acceptors, nutrients, pH, and inhibitory substrates or metabolites in bioremediation will be discussed with respect to the remediation of sites contaminated with either organics or heavy metals.

0-87371-613-2/94/$0.00+$.50
© 1994 by Lewis Publishers

One of the primary variables affecting the activity of heterotrophic bacteria is the ability and availability of reduced organic materials to serve as energy sources (Table 1). Whether a contaminant will serve as an effective energy source for an aerobic heterotrophic organism is a function of the average oxidation state of the carbon in the material. In general, higher oxidation states correspond to lower energy yields and thus provide less energetic incentive for a microorganism to degrade it. The oxidation level of chlorinated organic materials of concern as environmental contaminants is mainly a function of the number of halogen substituents. It should be noted that halogen substituents not only affect the redox level of the contaminant, but also the "recognition" of the substrate by microbial enzymes. This topic will be discussed in more detail in subsequent sections.

The outcome of each degradation process depends on microbial (e.g., biomass concentration, population diversity, and enzyme specificities), substrate (e.g., physicochemical characteristics, molecular structure, and concentration), and a range of environmental (e.g., pH, temperature, moisture content, E_h, availability of electron acceptors and carbon, and energy sources) parameters (Table 1). These parameters may affect an acclimation period of the microorganisms to the substrate or environment. The factors influencing the length of this acclimation period are outlined in Table 2. For a detailed description of the influence of these parameters on microbial degradation, the reader is referred to a recent excellent review.[2] The molecular structure and contaminant concentration have been shown to strongly affect the feasibility of bioremediation and the type of microbial transformation occurring, whether the compound will serve as a primary, secondary, or cometabolic substrate.[3–6]

2. ENVIRONMENTAL CHARACTERISTICS RELEVANT TO MICROBIAL ACTIVITY

One of the primary distinctions between surface soils, vadose zone soils, and groundwater sediments is the content of organic material. Surface soils, which typically receive regular inputs of organic material from plant shoots and roots, will therefore have a relatively high organic material content. The distinction regarding soil organic matter (SOM) is important in that high SOM levels are typically associated with high microbial numbers and a greater diversity of microbial populations. The organic matter also serves as a "storehouse" of carbon and energy as well as a source of other macronutrients such as nitrogen, phosphorus, and sulfur. Additionally, the relatively high negative charge of SOM electrostatically binds other mineral nutrients such as ammonium, calcium, potassium, and magnesium (e.g., cation exchange capacity). Subsurface soils and groundwater sediments characteristically have lower levels of organic matter and thus house lower microbial numbers and population diversity than surface soils. It should be mentioned that the above characteristics of surface soils may have been altered in many developed areas where topsoil has been stripped and replaced with fill.

Table 1.   Factors Affecting *In Situ* Biodegradation of Organic Pollutants

Physicochemical Factors
    Bio-unavailability of the pollutants by sorption processes
        Equilibrium sorption
        Irreversible sorption
        Incorporation into humic matter
    Mass transport limitations
        Oxygen diffusion and solubility
        Diffusion of nutrients
        Solubility/miscibility in/with water

Biological Factors
    Aerobic versus anaerobic processes
        Oxidation/reduction potential
        Availability of electron acceptors
        Microbial populations present
    Growth substrate versus cometabolism
        Type of contamination
        Concentration
        Alternative carbon sources present
        Microbial interactions (e.g., competition, succession)

Table 2.   Microbial, Environmental, and Substrate Factors Influencing Acclimation Period Proceeding Pollutant Degradation

Microbial
    Growth until "critical biomass" is reached
    Enzyme induction
    Mutation and gene rearrangement
    Enrichment of the capable microbial populations or plasmids

Environmental
    Depletion of preferential substrates
    Lack of nutrients
    Temporarily inhibitory environmental conditions

Substrate
    Too low concentration of pollutants
    Chemical structure of pollutants

Compared to undisturbed areas, developed sites can therefore be expected to have a lesser potential for *in situ* biodegradation activity.[7]

Although the diversity and numbers of microbial populations are a function of organic matter content, it cannot be inferred that microbial activity is nonexistent in deeper, low organic carbon (<0.1%) regions of the soil profile and in aquifers. Microbial numbers may decline several orders of magnitude with increasing depth, but have been shown to level off around 10^2 cells per gram soil.[8] In other cases however, microbial numbers have been reported to be independent of the examined depth profile.[9] It is difficult to discern a pattern in the changes, as the distribution and numbers of microorganisms are site-dependent and a function of the soil type.[10] In addition to a quantitative change, the predominance of different metabolic groups of microbiota changes with depth. Bacteria become more dominant in the microbial community with increasing depth in the soil profile, as numbers of other organisms such as actinomycetes and fungi decrease. This observation has been attributed to the ability of bacteria to utilize alternative electron acceptors to oxygen (Section 3).

Another important aspect of the physical composition of the environment on the occurrence and diversity of microbial populations concerns the texture of the surface or vadose zone soil and aquifer. Soil texture refers to the relative percentage of sand, silt, and clay, whereby a coarser material is defined as having a higher percentage of sand relative to clay. Key properties of the fine clay class are a high surface area and a higher cation exchange capacity (high negative charge). These factors combined yield higher numbers of microbial populations in clays relative to sands for similar reasons that SOM content influences the microbiota. However, in contrast to organic matter content, soil texture may vary independently of physical location. The texture of the soil in turn influences the water holding capacity or soil moisture content. Soil moisture content is one of the key factors controlling microbial activity in several respects. First, soil microorganisms are essentially aquatic life-forms and require at least thin films of water through which substrates and nutrients are assimilated and waste products excreted.[11] Second, soil moisture content and soil aeration are inversely related. Since moisture competes with gas for available pore space, an increase in soil moisture should be carefully considered as it might lead to depletion of oxygen. This condition will favor the activity of anaerobic organisms. Moisture level in this respect is irrelevant in groundwater sediments, which are characteristically saturated. In the latter case, the concentration of dissolved oxygen becomes a more relevant parameter in the assessment of the biodegradation potential.

Finally, microbial activity at a given location is influenced by atmosphere composition and temperature. The composition of the soil atmosphere relative to that on the surface differs primarily with respect to oxygen (up to 20%) and carbon dioxide (up to 0.5% versus 0.03% in the air) levels. The impediment to gaseous exchange presented by soils leads to a depletion in oxygen and possible enrichment of volatiles (e.g., carbon dioxide, methane, or low molecular weight organic acids) produced in the microbial decomposition of organic material. Temperature, on the other hand, influences microbial activity, as has been quantified by the Q_{10} coefficient, which evaluates the increase in microbial rate processes over $10°C$ intervals. Most biological systems have Q_{10} values of 1.5 to 3.0. These values apply in the temperature range where rate processes adhere to Arrhenius kinetics on the one hand and which are relevant for activity of soil microorganisms (generally between 10 and $50°C$) on the other hand. Above and below these ends of the spectrum, the effect of temperature on rate processes (Q_{10}) is proportionally much greater.

3. PHYSIOLOGY OF AEROBIC AND ANAEROBIC TRANSFORMATIONS

The basic processes and mechanisms of microbial aerobic and anaerobic metabolism are well understood and described in any textbook on the ecology and biochemistry of microorganisms.[12–14] The following is intended as a brief review of microbial physiology as it pertains to degradation of xenobiotic compounds. For more detailed information, the reader is referred to the above cited works or other chapters in this monograph. To understand the types, occurrence, and limitations of aerobic and anaerobic reactions, it is necessary to briefly examine the role of oxygen in the development of "ecological niches". Oxygen serves two functions in the degradation of organic matter: that of a terminal acceptor of electrons released during oxidation of organic carbon, and a reactant in processes mediated by oxygenases.

3.1 Types of Metabolic Processes

Primary metabolism of an organic compound has been defined as the use of the substrate as a source of carbon and energy for the microorganism resulting in the incorporation of organic carbon in cell matter. Thus, the substrate serves as an electron donor resulting in microbial growth. Secondary substrate metabolism occurs when the substrate can serve as a source of carbon and energy; however, the concentration is too low to enable a microbial population to grow or expand at its expense. Therefore, the degradation of low growth substrate concentrations either proceeds extremely slowly (since the biomass remains essentially unchanged) or has to be accomplished in the presence of another growth substrate (i.e., the primary substrate) at higher concentrations.

Application of cometabolism to site remediation of xenobiotics is required when the compound cannot serve as a source of carbon and energy by nature of the molecular structure, which does not induce the required catabolic enzymes, or simply the lack of enzymes able to mediate the required transformations. The term cometabolism has been defined as the metabolism of a compound that does not serve as a source of carbon and energy (i.e., no electron donor, only electron acceptor) or as an essential nutrient[15,16] and can only be achieved in the presence of a "primary (enzyme-inducing) substrate", appropriate electron donor, or by cells previously grown on an inducing substrate. Moreover, even if initial substrate transformations are achieved as the result of anaerobic cometabolism of aromatic compounds for example, the ring system has to be cleaved by the reducing bacterium to enable assimilation of the substrate carbon.[17] The requirement of electron donor and acceptor substrates in the previously described metabolic types is shown in Table 3.

3.2 Aerobic Processes

Aerobic transformations are characterized by metabolic processes involving oxygen as a reactant. Dioxygenases and monooxygenases are two of the primary enzymes employed by aerobic organisms during transformation and mineralization of xenobiotics. The functional difference between both enzymes is the number of atoms of molecular oxygen incorporated into the substrate. While both aromatic and aliphatic compounds may serve as substrates for monooxygenases, dioxygenase substrates are limited to aromatic compounds. The role of oxygenases

Table 3. Biological Processes and Environmental Conditions Under Which Organic Compounds May Be Transformed by Bacteria

Process	Examples	
	Halogenated	Non-halogenated
Primary Substrates		
Aerobic	Chlorobenzene, mono- and dichlorobenzoates (BA), mono-, di-, and pentachlorophenols, monochloro-biphenyls (BP), 2,4-D, 2,4,5-T, dibenzo-p-dioxin, dibenzofuran, chlorinated aliphatic acids	Benzene, benzoate, BTEX compounds, saturated and unsaturated hydrocarbons, naphthalene, creosote components
Anaerobic	Chlorobenzenes, dichlorophenols, dichloromethane	BTEX compounds, phenol, unsaturated aliphatic hydrocarbons
Cometabolism (Non-growth Substrates)		
Oxidations	Di- and trichloroBA, di- through pentachlorinated BPs, mono-, di-, and tetrachlorinated dioxins and furans, chloronaphthalene, trichloroethylene, dichloroethylene, vinyl chloride	Long-chain aliphatic hydrocarbons, branched chain aliphatics, some PAHs (anthracene, benzo[a]anthracene)
Reductions	Di-, tri-, tetra-, and hexachlorobenzenes, chlorobenzoates, chloroanilines, DDT, di- through heptachlorinated BPs, heptachlorinated dioxin, lindane, carbon tetrachloride, trichloroethane	Polycyclic aromatic hydrocarbons (PAH), aliphatic hydrocarbons

in xenobiotic degradation is perhaps more thoroughly investigated than that of any other class of enzymes and is reviewed elsewhere in this volume. A third class of oxygenases present in microorganisms is peroxidases, which to date have only been reported in lignin-degrading white rot fungi and derive their name from the fact that they have been shown to be peroxide-dependent. These enzymes have been receiving increasing attention for their role in xenobiotic degradation, in particular aromatic nuclei where they mediate single electron extractions, i.e., the formation of highly active radicals.

Other reactions pertinent to xenobiotic degradation, which are catalyzed by aerobes, but which have been shown to be independent of molecular oxygen, include dehalogenations mediated by hydrolytic dehalogenases. In this case, the halogen abstraction from the ring occurs via a nucleophilic substitution by the hydroxy-component of water in the medium. This reaction may thus occur under both aerobic and anaerobic conditions.

3.3 Anaerobic Processes

Anaerobic microorganisms take advantage of a range of electron acceptors, which, depending on their availability and the prevailing redox conditions, include nitrate (NO_3^-), iron and manganese, sulfate (SO_4^{2-}), or carbon dioxide. Hence, the ecological classification results in nitrate reducing (denitrifying), iron and manganese reducing, sulfate reducing, and fermentative or methanogenic microorganisms.[12,18] The range of redox potentials at which each ecological class of organisms is active is presented in Figure 1 (modified from Zehnder and Stumm[18]). In the open environment, this translates into a stratification of microbiota (Figure 1). Due to the poor diffusion of oxygen in soil and water (especially stagnant bodies) and the structural heterogeneity of soils, the oxygen-dominated environments are limited to surficial soils. Even in these areas, anaerobic microenvironments may be present due to the biological depletion of the oxygen and restricted diffusion of the overlying atmosphere into soil pore spaces. The oxygen concentration of sediment environments is influenced by the conditions of the water column above. Hence, a succession of the different anaerobic physiological groups is generally observable with sediment depth. While methanogenesis and sulfate reduction have been reported to occur simultaneously,[19-22] nitrate-reducing microorganisms are likely to be found at shallower depths (higher redox potentials) than both of the previous groups.[18,23-25] Thus, denitrification, sulfate reduction, and fermentation coupled to methanogenesis may occur concurrently in the same soil or sediment, as different conditions exist in different microhabitats.[12]

The growth yield of anaerobic bacteria is extremely low due to low energy yields (biochemically expressed as mol adenosine triphosphate (ATP) formed/mol substrate consumed, or chemically expressed as the available free energy, $\Delta G^{\circ\prime}$ in kJ/mol substrate, obtained from any given reaction) during anaerobic metabolism.[18] Thus, usually less than 10% of the carbon from a completely mineralized substrate can be incorporated into cell matter, in contrast to almost 50% in the case of aerobic bacteria. However, during metabolism of xenobiotic organic compounds, the

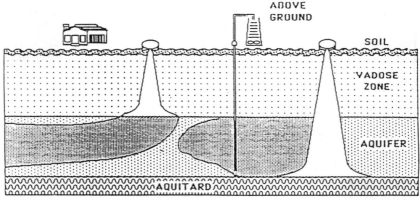

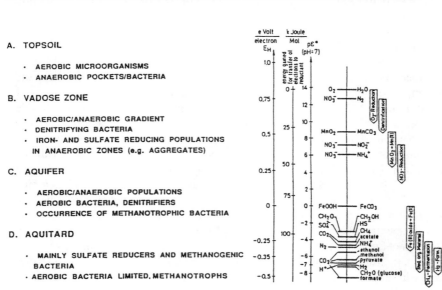

Figure 1.   Different locations of soil contamination and likely ecological distribution of bacterial populations.

enzymes involved are often unable to catalyze complete degradation of the substrate, rather only initial transformations such as oxidations or decarboxylations occur. Hence, the bacterium is unable to assimilate carbon or obtain energy from the substrate. Alternatively, nearly every substrate, including glucose, supplied at sufficiently low concentrations will cover only the maintenance energy needs of a microbial population, and growth is possible only beyond a threshold substrate concentration.[6,26] This is often the case with xenobiotics in the environment, which are present in very low (ppm or lower) concentrations.

The range of degradative reactions possible in anaerobic environments is rather restricted and includes the following: hydrogenations, dehydrogenations, hydrations, dehydrations, hydrolyses, condensations, and photoreactions. Also impor-

Table 4.   Anaerobic Microbial Transformation Reactions of Organic Compounds

Reaction Type	Examples
Hydrogenations/Dehydrogenations	
$ArOH + 6H^+ + 6e^- = C_6H_{11}OH$	Phenol, catechol, benzoate, benzene
$R-OH = (R-OH)^{2+} + 2H^+ + 2e^-$	Unsaturated hydrocarbons and fatty acids
Hydrations/Dehydrations	
$(R-OH)^{2+} + H_2O = R-COOH + 2H^+ + 2e^-$	Unsaturated hydrocarbons and fatty acids
$Ar(OH)_2 + 2H^+ + 2e^- = ArOH + H_2O$	Hydroquinones, hydroxybenzoates
Hydrolyses	
$R-ArX + H_2O = R-ArOH + HX$	Halogenated benzoates, phenols
$RX + H_2O = ROH + HX$	Halogenated alkyl chains
Carboxylations/Decarboxylations	
$R-ArOH + CO_2 = R-ArOH-COOH$	Cresol, toluene, short hydrocarbons
$Ar(OH)_2-COOH = Ar(OH)_2 + CO_2$	Benzoate, hydroxybenzoates, phthalates
Reductive Dehalogenation	
Hydrogenolysis:	
$RX + H^+ + 2e^- = RH + X^-$	Chlorinated solvents, polychlorinated aromatic compounds
Dihalo-elimination:	
$CX-CX + 2e^- = C{=}C + 2X^-$	Hexachloroethane
Coupling:	
$2RX + 2e^- = R-R + 2X^-$	Tetrachloromethane
Coenzyme B_{12}-Dependent Reactions	
Demethylation:	
$R-O-CH_3 + B_{12} = R-OH + B_{12}(CH_3)$	Alkyl methyl ethers
$R-O-ROH + B_{12} = R-O-ROH + B_{12}(CH_3)$	Polyethylene glycol
Dechlorination:	
$RCl_4 + H^+ + e^- + B_{12} = RCl_3 + B_{12} + Cl^-$	Chlorinated ethylenes, chloromethanes
$ArCl_5 + H^+ + e^- + B_{12} = ArCl_4 + B_{12} + Cl^-$	Pentachlorobenzene, -phenols, 2,4,5–T, PCBs
Methylation:	
$Hg^{2+} + 2CH_3^- + B_{12} = Hg(CH_3)_2 + B_{12}$	Heavy metals

Source: Vogel et al.[27]

tant are carboxylations, decarboxylations, and coenzyme B_{12}-mediated reactions.[26] A separate class of anaerobic reactions deals with the phenomenon of reductive dehalogenation, which is imperative prior to further metabolism mediated by the reactions mentioned earlier. The application of these reactions pertaining to the transformation and degradation of different classes of anthropogenic compounds in controlled laboratory environments and in the field will be described in a subsequent section. An exemplary summary of the reactions described is presented in Table 4 (modified from Vogel et al.[27]).

Although the concept of reductive dehalogenation dates back to the mid-1960s for alkyl halides,[28] dehalogenation of aromatic compounds has not been reported until almost two decades later.[29] Since then, reductive dehalogenation of chlorinated aliphatic compounds, solvents, and aromatics has been demonstrated to occur under anaerobic biotic (mainly methanogenic) or abiotic (e.g., in the presence of vitamin B_{12}, reduced Fe (II), and reduced iron-porfyrins) conditions.[30,31]

Redox proteins and biochemical cofactors such as cytochromes, corrinoids, and flavins are responsible for electron transfer within microorganisms. Mechanistically, the reduction process is believed to involve electron transfers and the formation of a radical, which is then hydrogenated; two-electron transfer has been shown to occur less frequently (Table 4).[32] Hence, if an electron-transfer macromolecule has electrons of sufficient energy as well as a structure permitting release of those electrons to halogenated compounds, reductive dehalogenation can occur.[33] The subject has recently been reviewed.[2,33-35] With few exceptions, anaerobic transformation of halogenated compounds is initiated by the removal of the bulky halogens by reductive dehalogenation. Reduction must be coupled to the oxidation of an electron donor. In some cases such as with dichlorophenols,[36] the compound under study can serve both as an alternative electron "sink" as well as the electron donor. In contrast, compounds such as the chloroethenes require exogenous electron donors to supply reducing equivalents. Thus, reductive dehalogenation depends, at least in part, upon the electron donor requirements and the efficiency with which electrons can be directed to dehalogenation.[33] The electron transfer efficiency, in turn, is highly (but not solely) dependent on the half reaction reduction potential ($\Delta G^{\circ\prime}$) of the reductant (i.e., the halogenated substrate) and that of the electron transfer system (i.e., the biochemical macromolecules in denitrifying, sulfate-reducing, and methanogenic systems). Analogously, the electrons used for abiotic reductions originate in the bonds of organic or inorganic compounds and are transferred to halogenated compounds provided the electrons are of sufficient energy and a structure permitting electron transfer is present. Overall, these reductive processes are well suited for highly halogenated compounds deficient in electrons and hydrogen, where hydrolytic and oxidative processes are better suited for the destruction of less halogenated compounds relatively rich in electrons and hydrogen.[33] The latter situation is described in the section on aerobic processes.

Until recently, reductive dehalogenation was generally considered a detoxification mechanism which does not yield energy. Hence, reductive dehalogenation can be regarded as a form of "anaerobic cometabolism", whether or not the compound is further degraded. A well-documented exception is that of *Desulfomonile tiedjei* gen. nov. sp. nov., which was identified in an anaerobic 3-chlorobenzoate-degrading consortium to produce ATP (high-energy storage molecules) from the dechlorination step, relative to controls.[37,38]

4. PHYSIOLOGICAL CONSIDERATIONS OF SPECIFIC TRANSFORMATIONS IN THE ENVIRONMENT

4.1 Aliphatic Hydrocarbons

Microbial uptake of aliphatic hydrocarbons is severely impeded by their relatively low aqueous solubility.[39,40] Microbial adaptations for hydrocarbon uptake include excretion of emulsifiants, which capture hydrocarbon droplets (microdroplet smaller than the microbial cell or macrodroplets) within a surfactant micelle. The formation and physiological role of biosurfactants has recently been reviewed.[41] The appearance of biosurfactants in the culture medium or attached to the cell boundaries is often regarded as a prerequisite for initiation of biodegradation.[42] However, addition of purified biosurfactants resulted in inhibitory as well as in stimulatory effects on growth.[43,44] Addition of sophoroside strongly inhibited growth on hexadecane by *Acinetobacter calcoaceticus*,[44] while rhamnolipids reportedly stimulated hydrocarbon metabolism by *Pseudomonas aeruginosa*,[45,46] and sophorose lipids stimulated metabolism of a number of alkanes by strain KSM-36.[47]

Conversions of saturated hydrocarbons to methane and carbon dioxide by methanogens or anaerobic oxidation of methane with sulfate as an electron acceptor in the presence of acetate have been inferred indirectly by thermodynamic calculations[48,49] or postulated on the basis of geochemical data.[50] In the latter case, methane would serve as the electron donor, sulfate as the electron acceptor, and acetate as the carbon source. However, no conclusive evidence exists to substantiate these claims. Further complications are that the reports on anaerobic degradation of saturated hydrocarbons did not deal with completely anoxic cultures. Microbial degradation of aliphatic hydrocarbons is therefore largely dependent on molecular oxygen.[26,39,51,52] Application of bioremediation as a process for removing these contaminants from low-oxygen environments (e.g., groundwaters, sediments) is therefore likely to meet with limited success.

Anaerobic metabolism of unsaturated hydrocarbons with at least one double bond has been postulated to proceed via a hydration of the double bond to form an alcohol. Both shorter and longer length hydrocarbons as well as branched hydrocarbons are less effectively and incompletely degraded.[26] Anaerobic metabolism of unsaturated aliphatic hydrocarbons has so far been demonstrated under nitrate-reducing,[53] sulfate-reducing,[54] and methanogenic[26,55] conditions. However, although the latter is an energetically favorable process ($\Delta G^{\circ\prime} = -102$ kJ $\cdot$ mol ethene[-1], $\Delta G^{\circ\prime} = -208.7$ kJ $\cdot$ mol acetylene[-1]), no suitable enzyme systems to bring about these reactions have been observed. In general, aerobic degradation is preferable for remediation of both saturated and unsaturated aliphatics, although unsaturated aliphatics are amenable to anaerobic degradation processes.[56]

4.2 Aromatic Hydrocarbons

The information on the aerobic degradation of nonchlorinated aromatic hydrocarbons (AH) is vast and is discussed elsewhere in this volume. For further information on heterotrophic metabolism of AH, the reader is referred to existing reviews on the matter.[57,58] The majority of information concerning anaerobic degradation of aromatics is derived from studies with benzoate and has been reviewed recently.[2,25,59,60] The present discussion will focus on the types of reactions responsible for degradation under different redox conditions.

The reductive pathway established for benzoate has been demonstrated under methanogenic, nitrate-reducing, and photosynthetic conditions.[25,61,62] Reduction of the aromatic ring results in the formation of 2-oxocyclohexane carboxylic acid prior to ring fission. A sequence of hydroxylation and dehydrogenation reactions, analogous to β-oxidation of fatty acids, is thought to occur prior to adipate (C_4-acid) formation, which could then be further used as a growth substrate. The first evidence of benzoate metabolism under sulfate-reducing conditions was reported by Widdel.[63] Methanogenic bacteria have been described, which either incompletely (i.e., without ring fission) reduce and decarboxylate or completely (i.e., with ring fission) degrade benzoate. Proposed pathways include reductive cleavage of the ring to a C_7 carboxylic acid (heptanoic acid) or a C_7 dicarboxylic acid, which is readily available to methanogenic bacteria.

Another group of compounds, which has received increasing attention because of their confirmed or suspected carcinogenicity,[64] are benzene, toluene, ethylbenzene, and the xylene isomers (BTEX). While BTEX degradation proceeds rapidly and has been widely described under aerobic conditions (elsewhere in this volume), anaerobic transformation is not well understood and is a recent phenomenon. The earliest report on anaerobic degradation of a BTEX compound, benzene, dates back to 1980.[65] Since the mid-1980s, laboratory studies have shown that the anaerobic degradation of BTEX compounds can occur over the whole range of redox potentials. Nitrate-reducers,[66–72] sulfate-reducers,[72–75] iron-reducers,[76,77] and methanogenic consortia[78–80] have all been demonstrated to metabolize at least one BTEX compound. While the range of BTEX degradation spans the whole spectrum of redox conditions, a rather large degree of variability has been reported with respect to conditions allowing metabolism of the individual components. This implies that not all BTEX compounds will necessarily be metabolized under the same redox conditions and in the same sequence at a given site. Toluene has most frequently been reported to be the first BTEX component to be degraded, followed by the xylenes. o-Xylene appears to be the most recalcitrant isomer; however, the pattern of isomer degradation is site specific. The transformation of ethylbenzene was described to occur under methanogenic and denitrification conditions.[78,81]

Degradation of BTEX under specific redox conditions has been demonstrated via stoichiometrical coupling of substrate disappearance and electron-acceptor concentrations[66,72,78,82] or via the appearance of ^{14}C-labeled carbon dioxide.[74,75] Whereas the limited number of metabolic intermediates identified may suggest that hydroxylation, dehydrogenation, demethoxylation, and decarboxylation reactions feed these compounds into the reductive benzoate pathway described earlier,[58,77] the actual pathways of anaerobic metabolism of BTEX are largely unknown. Roberts et al.[83] and Ramanand and Suflita[72] demonstrated a novel carboxylation reaction during m-cresol degradation by a mixed methanogenic culture yielding 4-hydroxy-2-methylbenzoic acid as an intermediate, which was further mineralized. Bisaillon et al. showed a similar carboxylation reaction during anaerobic metabolism of o-cresol.[84] Recently, Evans et al. proposed a novel degradation pathway for toluene under denitrifying conditions involving a nucleophilic attack by acetyl-CoA proceeding via phenylpropionyl-CoA and benzoyl-CoA.[85] It appears this pathway may be more widespread among different physiological groups, as it has recently been suggested to occur under sulfate-reducing conditions as well.[86]

Under aerobic conditions, both beneficial and detrimental substrate interactions have been invoked in complete degradation of all BTEX compounds. Alvarez and Vogel found that whereas benzene and p-xylene degradation by a pure culture isolated from sandy aquifer material was dependent on the presence of toluene, the presence of p-xylene retarded benzene and toluene degradation.[87] o-Xylene is usually reported to be the most recalcitrant of the BTEX compounds,[88,89] although some pure cultures capable of using o-xylene as a growth substrate have been isolated.[90]

Anaerobic degradation of aromatic compounds with methoxy- and acetyl-sidechains such as the humic components ferulic acid, vannilic acid, and syringaldehyde have been shown to proceed via a series of hydrations, decarboxylations, and acetate removal into the benzoate degradation pathway.[2,25,91]

The anaerobic degradation of polycyclic aromatic hydrocarbons (PAH) has only recently been described for naphthalene and acenaphtene in water saturated microcosms and under denitrifying conditions, although no metabolites were determined.[92,93] Respectively 2-ethylphenol[94] and p-cresol[95] were identified as metabolites from naphtalene degradation in methanogenic microcosms presumably as the result from ring reduction and reductive ring fission. Biphenyl degradation has only been demonstrated under aerobic conditions, except for 2-hydroxybiphenyl. Suflita et al.[96] showed that aquifer-derived inoculum degraded the substituted biphenyl under sulfate-reducing conditions, as demonstrated by electron donor/acceptor stoichiometry. However, metabolism of 2-hydroxybiphenyl was isomer specific; neither meta- nor para-hydroxylated biphenyls could be degraded.

4.3 Heterocyclic and Alicyclic Compounds

Heterocyclic compounds are chemically more susceptible to anaerobic oxidation than homocyclic aromatics. Because the heteroatom distorts the electron density in the ring structure, protonation (ring activation) and nucleophilic attack by water are facilitated.[97,98] The earliest microbial oxidation products detected in methanogenic and fermentative incubations with most nitrogen-containing heterocyclics were hydroxylated analogs of the added substrates, which in water tautomerized to oxo-isomers.[99,100] These anaerobic oxidations are biologically initiated by dehydrogenases, after which aqueous oxygen is incorporated preferably into the nitrogen-containing ring, as was demonstrated with $H_2^{18}O$ experiments with quinoline.[101] Anaerobic oxidation of s-triazines has shown to replace the nitro-substituent on melamine (trinitro-triazine).[102]

Transformation of nitrogen-containing heterocyclics has also been demonstrated with denitrifying[103] and sulfate-reducing[104] enrichments and microcosms. Alternatively, bacterial degradation products of acridine (a three-membered, fused ring containing one heterocyclic) have been detected under methanogenic conditions[105] presumably after initial hydroxylation of either the heterocyclic or aromatic ring.

Degradation of sulfur-heterocyclics has been reported to occur under methanogenic or fermentative conditions.[106,107] The first intermediate in benzothiophene degradation has been reported to be oxobenzothiophene time as well as and mono-aromatic degradation products presumably resulting from hydrolytic ring cleavage. The appearance of aliphatic products suggests mineralization of the sulfur-heterocyclic, which serves as a source of sulfur.

4.4 Alkyl Halides

This group of compounds includes, but is not restricted to, chlorinated pesticides (e.g., 1,1-bis-4-chlorophenyl-2,2,2-trichloroethane or DDT, hexachlorocyclohexane or lindane) and halogenated alkanes and alkenes (e.g., chlorinated solvents such as trichloroethylene and tetrachloromethane). The widespread use of DDT throughout the last three decades and its extreme persistence have made DDT one of the most widely studied pesticides.[28,108–112] Halogenated alkanes and alkenes have received considerable attention, as chlorinated solvents have become the most frequently detected groundwater contaminants.[113]

Aerobically, DDT has been reported to be cometabolically degraded by both bacteria[108,109] and fungi.[114] Khöler et al.[114] reported that 90% of the DDT added was removed from cultures containing the white rot fungus *Phanerochaete chrysosporium*. While the metabolic mechanisms were not fully elucidated, it was clear that removal proceeded even in the absence of detectable lignin peroxidase activity. Bacterial degradation of these chemicals was presumably mediated by enzymes. In either case, the maintenance of degradative activity appeared to be dependent on a supply of a readily available energy source. Since no specific inducer substrates were necessary, degradation apparently hinged on the presence of a specifically adapted (or preexposed) population.

Anaerobically, dehydrochlorination to 2,2-bis-4-chlorophenyl-1,1-dichloro-ethylene (DDE) and reductive dechlorination to 2,2-bis(4-chlorophenyl)-1,1-di-chloroethane (DDD) appear to be the dominant degradation mechanisms presumably as enzymatic detoxification responses. Some microorganisms have been reported to bring about further dechlorination and degradation.[111] Many other chlorinated insecticides have been shown to undergo reductive transformation processes. Whether or not these include dechlorinations has not been demonstrated.[28,115] Lindane (γ-hexachlorocyclohexane) was reductively dechlorinated to tetrachlorocyclohexane under anaerobic conditions such as exist in flooded rice soils.[116] Similar reductive processes of DDT were reported with reduced iron-porfyrins[117] or reduced (Fe^{2+}) cytochrome oxidase indicating that for example hematin can catalyze reductive dehalogenation under low redox potentials.[35] Additionally, porfyrins and corrins have been shown to catalyze reductive dechlorination of lindane,[118] Mirex,[119] and toxaphene.[120]

The susceptibility of nematocides such as ethylenedibromide and 1,2-dibromo-3-chloropropane to reductive dehalogenation had been shown two decades earlier.[121] Recently, di-, tri-, and tetrahalomethanes and ethylenes have been shown to be reductively dehalogenated by methanogenic bacteria in bioreactor studies.[27,33,122] Alternatively, haloalkanes and chloromethanes have been shown to be dehalogenated and/or mineralized by methylotrophic,[123–125] denitrifying,[126] and anaerobic hydrocarbon-utilizing[127] bacteria. Instead of a reductive dehalogenation mechanism, respectively dehalogenases and halidohydrolases are thought to be responsible for chlorine removal. Dehalogenases catalyze hydrolysis of halogen-substituted alkanoic acids yielding either hydroxyalkanoic acids from monohalogenated acids or oxoalkanoic acids from dihalogenated acids,[128] products that can easily be further metabolized as outlined earlier (Section 4.1). The different mechanistic reactions are presented in Table 4.

The *in situ* transformation of tetrachloromethane (or carbon tetrachloride, CT), 1,1,1-trichloroethane (TCA), and two chlorofluorocarbons was evaluated under anoxic conditions in a shallow, confined aquifer (Moffett Field Naval Air Station, California). Biostimulation was accomplished through introduction of acetate as the electron donor and nitrate as the electron acceptor.[129] Although all compounds were transformed, the rate of CT disappearance was higher and commenced 2 weeks after active denitrification began. Both chloroform and CO_2 were observed as CT degradation intermediates suggesting that both denitrifiers and an unknown secondary population (presumably aerobic) were responsible for the biotransformation. Anaerobic reductive dechlorination of compounds such as perchloroethylene (PCE) indicated that a potential treatment process could involve *in situ* biostimulation of methanogens. However, the sequential dehalogenation resulted in the transient accumulation of vinyl chloride, which has proven to be a potent carcinogen.[130] For a list of halogenated aliphatic compounds transformed by anaerobic microorganisms, the reader is referred to Vogel et al.[27]

Chlorinated ethenes have been found to be cometabolized aerobically by methanotrophic microorganisms. This physiological type of bacteria thrives on the interface between aerobic and anaerobic environments in the presence of stable sources of methane such as those generated by methanogens. These organisms harbor broad-spectrum methane-monooxygenases (MMO), which are central to the methane-oxidizing pathway and require reducing power for activity.[131] Methanotrophic bacteria have been shown to oxidize compounds containing unsaturated carbon-carbon bonds such as di- and trichloroethylene (DCE, TCE) and styrene,[132–134] and aromatic ring structures such as naphthalene and lesser chlorinated or hydroxylated biphenyls.[135] Thus degradation of compounds such as PCE by methanotrophs has not been demonstrated.

Because of their versatility, biostimulation of methanotrophs *in situ* for remediation of contaminated aquifers has been investigated. Field studies on TCE, *cis*- and *trans*-1,2-DCE, and vinyl chloride (VC) transformation at Moffett Field Naval Air Station (California) have demonstrated the potential of methanotrophs for *in situ* aquifer restoration[136,137] and have shown that inhibitory substrate interactions demonstrated in the laboratory apply in the environment as well.[138] Whereas methane competes with TCE for the MMO and thus interferes with TCE degradation, the generation of epoxides and carbon monoxide intermediates may have inhibitory effects as well once certain concentrations are reached.[139–141] The essential features of the process include the introduction of oxygen to provide an environment suitable for the aerobic methanotrophs and a carbon source (e.g., methane, phenol) to support growth and the generation of reducing power and to sustain enzyme activity.[142] The applicability of these results is now being evaluated in a TCE-contaminated groundwater plume in St. Joseph, Michigan.[143] Issues that must be addressed are the supply of reductant and maintaining activity of the biomass. The former may be addressed by the substitution of formate for methane; however, formate does not induce the MMO.

Summarizing, under anaerobic conditions, halogenated aliphatic compounds often serve only as alternative electron acceptors utilized by methanogenic bacteria. In these cases, the bacteria grow at the expense of other easily metabolizable, energy-generating substrates such as short-chain aliphatic and aromatic acids and cometabolize the alkyl halides. In the situation where the microorganism possesses dehalogenases or halidohydrolases, the halogenated compound itself, provided it is present at concentrations enabling sustenance of a microbial population, usually serves as a source of carbon and energy.

4.5 Aryl Halides

Probably the most diverse group of environmental contaminants, aryl halides (chlorinated aromatic hydrocarbons, CAH) have gained notoriety because of their recalcitrance towards biodegradation and to anaerobic transformations in particular.[34,144,145] While the simplest six-membered halogenated carbon cycle is chlorobenzene, the structural diversity of CAHs ranges from one to six chlorines per ring to multiple fused (e.g., chloronaphthalenes), carbon-linked (e.g., polychlorinated biphenyls), or ether-linked (e.g., chlorodibenzo-*p*-dioxins, chlorinated furans, diphenyl ethers) aromatic rings. For a summary of the molecular structures discussed, the reader is referred to Rochkind-Dubinsky et al.[34] The hydrophobic nature of aryl halides contributes to their limited transport in soil and thus their concentration in the top layers with prevailing aerobic conditions, where they are strongly sorbed to soil organic matter (SOM) or have been irreversibly incorporated — thus rendered biologically unavailable — into humus.[2,146–148] Transport of aryl halides has been described to be associated with particulate matter or aerosols, which may account for their wide distribution in anaerobic river and lake sediments.[149]

Since Suflita et al.[29] described the anaerobic reductive dehalogenation of aromatic iodo-, bromo-, fluoro-, and chlorobenzoate compounds, additional information has emerged to substantiate its importance as a biotransformation mechanism in the anaerobic environment. Simple halogenated aromatic hydrocarbons are reductively dechlorinated provided they have a minimum of two chlorine substituents. Hexachlorobenzene was sequentially reduced to di- and trichlorobenzenes in batch microcosms using anaerobic sewage sludge,[150,151] while Bosma et al.[152] demonstrated reductive dechlorination of all trichlorobenzene isomers to monochlorobenzene in columns containing anaerobic sediments from the Rhine River. In all cases, simple aliphatic or aromatic acids were provided as electron donors for the microbial populations in the methanogenic sediments, while the chlorobenzenes served as "coreductive" electron acceptors, next to carbon dioxide. Alternatively, pentachlorophenol dehalogenation has long been observed in rice paddy soils and has more recently been demonstrated in laboratory microcosms.[153–156] In these studies, chlorophenols were provided as the electron donors, which enhanced the rates of dechlorination compared to controls. The pattern of chlorine removal depends on the microbial population involved, although the rate of *meta*-chlorine removal appeared to be the slowest.

The reductive dechlorination of polychlorinated biphenyls (PCBs) was first reported *in situ* in anaerobic river sediments.[157] When chromatograms of original Aroclor™ mixtures were compared with those of the sediment samples, the authors showed that the highly chlorinated congeners (three to six chlorines) had decreased significantly with a concurrent increase in mono-, di-, and some trichlorobiphenyls. Depending on the microbial population involved, either *ortho*- or *meta*- and *para*-substituted chlorines were removed under methanogenic conditions. Quensen et al.[158] demonstrated in laboratory microcosms using Hudson River (New York) sediment that Aroclor™ 1242 (average of four chlorines), Aroclor™ 1254 (average of five chlorines), and Aroclor™ 1260 (average of six chlorines) all dehalogenated according to the pattern described above, the latter albeit slowly. Investigations on the influence of the type of growth substrate (i.e., electron donor) on the rate of dechlorination revealed that methanol and glucose were better substrates than acetone and acetate.[159] Samples not amended with organic carbon, yet supplied with the same nutrient concentrations as the aforementioned, did not affect dechlorination within the time monitored. Ye et al.[160] recently reported on the influence of heat and ethanol treatment on anaerobic dechlorination of PCBs. Their results indicated that treated cultures retained their dechlorinating activity, albeit less extensive than untreated controls. Treated cultures preferentially removed *meta*-chlorines compared to both *ortho*- and *meta*-dechlorination in the controls. Recently, Adriaens and Grbić-Galić[161] provided the first evidence that highly chlorinated dibenzo-*p*-dioxins (PCDDs) and dibenzofurans (PCDF) may undergo similar reductive dehalogenations in methanogenic microcosms inoculated with Hudson River sediment or with dichlorophenol enriched cultures derived from a waste pond (Cherokee Pond, Georgia). Small amounts of heptachlorinated dioxin were converted to a hexachlorinated congener.

Similar dehalogenations of a pentachlorinated biphenyl[31] and chlorobenzenes[30,31] have been demonstrated using vitamin B_{12} or porphyrins and corrinoids. Single chlorines in either *meta*- or *para*-position were removed.

As indicated above, some aryl halides may serve as growth substrates under anaerobic conditions. Dehalogenation appears to be the rate-limiting step during biotransformation, as the aromatic ring can only be further degraded via the benzoate pathway described earlier once all the halogen substituents are removed. Amendment of the inocula with easily metabolizable organic substrates (i.e., electron donors) is essential for reductive dehalogenation processes. The most effective substrate is apparently dependent on the microbial population involved, rather than on the chemical structure of the compound to be reduced.

Aerobically, the degradation of CAHs is a lot more complex and is largely dependent on "substrate recognition" by the metabolic enzymes, mono- and dioxygenases in particular. Hence, cometabolic substrate interactions are often a prerequisite during biotransformations of CAHs. Highly chlorinated CAHs are already very oxidized and will therefore not likely be susceptible to further biological oxidation. As the range of CAH used as a growth substrate or cometabolized by aerobic microorganisms is rather extensive (Table 3), the reader is referred to other sources for a more detailed review.[34,162] Rather, the potential and physiology of aerobic CAH degradation will be discussed using environmentally pertinent examples.

The aerobic degradation of CAH can be divided in three types of metabolic processes: (1) the compound is able to serve as a sole source of carbon and energy to a single organism or a sequence of microorganisms; (2) the compound has to be degraded via cometabolic processes by virtue of the lack of "recognition" by the enzymes involved and thus serves not as a growth substrate; or (3) the product formed from initial cometabolic transformations can serve as a growth substrate for either the same or a different microorganism. Obviously, the first type would be the most favorable for bioremediation purposes. The impediments presented by acclimation to and concentration of the compound (as referred to earlier) notwithstanding, CAHs such as chlorobenzenes and benzoates, chlorophenols, and some chlorinated pesticides (e.g., phenoxyacetic acids), have been reported to serve as primary growth substrates. Two main pathways have been described: either the ring is dehalogenated prior to ring cleavage or after ring fission. In the first case, hydrolytic dehalogenases replace the halogen by a hydroxyl from water. Aromatic monooxygenases then oxidize the ring to form a dihydroxy intermediate, which can then be further cleaved by dioxygenases, as is the case with for example chlorobenzoates and chlorophenols.[163–165] Some of the chlorines on highly chlorinated CAH such as pentachlorophenol may be reductively removed, even under aerobic conditions.[166] Alternatively, after initial oxidation by dioxygenases to form chlorinated dihydrodiols, the ring can be cleaved either between or adjacent to either hydroxyl (ortho- or meta-ring fission) by ring cleaving dioxygenases and dehalogenated (e.g., chlorobenzenes, chlorocatechols).[162,167,168] In the case of PCBs, cometabolic processes are required to degrade the compound. Analog enrichment constitutes a special case of cometabolism whereby the nonchlorinated analog (which is a natural substrate for the enzyme) is provided as an enzyme-inducing substrate. Due to the nonspecificity of the enzyme, the CAH will be recognized by the enzyme and fortuitously transformed. To enhance aerobic PCB transformation, biphenyl is added to induce for the biphenyl dioxygenases.[149,169–171] After cleavage of one of the rings (usually the lesser chlorinated ring), mainly chlorobenzoates accumulate, which can then serve as growth substrates for different microorganisms.[165,172] Cometabolic processes may also be required for other (usually higher chlorinated) CAHs depending on the microorganism involved. Although PCB degradation constitutes a case where the product from cometabolism may serve as a growth substrate for commensals, examples have been described where the same organism is responsible for the initial cometabolic step(s) and further degradation of the CAH.[173,174]

Aerobic transformations of CAH and polycyclic aromatic hydrocarbons (PAH) by fungi have been documented to span a wide range of substrates. The most versatile genera appear to belong to the white rot fungi *Phanerochaete* sp., which are able to mediate hydroxylations, hydrations, and dehalogenations, as has been demonstrated with chlorobenzenes, pentachlorophenol, PCBs, and DDT.[175,176] The nonspecific hydrogen peroxide-inducible lignin peroxidases fortuitously oxidize CAH and PAH due to the similarities in molecular structure to lignin components.[177] Intermediates observed include, aside from carbon dioxide, arene oxides (epoxides), dihydrodiols, and monohydroxylated analogs. In a small scale field study, Lamar and Dietrich[178] used infested wood chips (which have shown to absorb PCP) to amend a plot contaminated with pentachlorophenol (PCP). After an initial lag period, 90% of the PCP added disappeared after a 6.5 week incubation under suboptimal growth temperatures and without the addition of inorganic nutrients, relative to noninoculated chips. A small amount of PCP decrease was the result of fungal methylation to pentachloroanisole. Moreover, examples of fungi belonging to the genera *Aspergillus*, *Trichoderma*, and *Penicillium* have been demonstrated to mediate transformations and mineralizations (CO_2 evolution) of CAHs. In these cases however, the main intermediates observed from mineralization are methylated analogs (i.e., anisoles in the case of chlorophenols), except for in the case of chlorinated phenoxy herbicides (2,4-D, 2,4,5-T) where hydroxylation occurred.[34]

Due to the vast number of intermediates formed from aerobic transformation and mineralization of CAH, potential problems arise when aerobic processes are considered for bioremediation of CAH. Whereas Alexander[15] has drawn attention to the formation of more toxic intermediates, the formation of more mobile products has to be considered with regards to groundwater contamination. Such would be the case with PCB cometabolism where more mobile chlorobenzoates are formed. Moreover, inhibitory substrate interactions between product, cometabolic substrate, and growth substrate may present further impediments to complete mineralization of the CAH.[167,179,180] From an engineering point of view, there remains the problem of maintaining the contaminated site sufficiently aerobic as to not inhibit the microorganisms responsible for the desired degradation. Amendment with hydrogen peroxide is a much favored and effective technique.[34]

5. PHYSIOLOGY OF DETOXIFICATION MECHANISMS OF HEAVY METALS AND ORGANO-METAL COMPLEXES

Heavy metals in the environment have been in the forefront of academic and regulatory concern due to their highly toxic nature and translocation through the food chain. Their speciation and biogeochemistry are highly dependent on the acidity of the environment[181] and the activity or presence of microorganisms capable of heavy metal transformations,[182,183] whether as elemental species or organic ligands (e.g., metal-cyano complex, methylated species, ...). Microorganisms have developed transformation mechanisms to detoxify the abundance of heavy metals because of (1) their competition with biochemically important elements such as sulfur and phosphorus (e.g., selenium and arsenic), (2) their solubility and affinity for cell membranes (e.g., mercury, chromium), or (3) their general metabolically inhibitory effects (e.g., cadmium, ...). The physiological adaptations and detoxification mechanisms of heavy metals employed by microorganisms include (1) oxidations and reductions (e.g., mercury, selenium, chromium, arsenic, ...), (2) immobilization and mineralization (i.e., microbial processes that influence transformations between inorganic and organic forms of an element), and (3) methylation and/or volatilization. The methylation of heavy metals is an important process because it often leads to a change in both mobility and toxicity. Microorganisms have been shown to be important catalysts in the methylation of mercury,[183–186] arsenic,[187,188] tellurium,[189] selenium,[190,191] tin,[192] and lead.[193]

As it is not within the scope of this chapter to review the biological transformations of manganese and iron, which may serve as alternative electron acceptors to microorganisms under reduced conditions ($E_h = -0.1$ to 0.5 V, Figure 1), the reader is referred to excellent texts by Ghiorse[194,195] and Lovley.[196] Hence, the physiology of possible transformation mechanisms of mercury, selenium, cadmium, arsenicum, chromium, free cyanide, organo-cyanide complexes, uranium, and cesium under different environmental conditions is described. Moreover, some attention will be devoted to the importance of microbial ligands and exopolymers in the mobilization and availability of heavy metals in general.

5.1 Bacterial Exopolymers and Metal Mobilization

Microorganisms produce and excrete organic ligands, which are directly involved in the mobilization of metals.[197] Low molecular weight organic acids, bacterial siderophores, and larger bacterial exopolymers engage in complexation and chemical binding reactions with metals.[198] The organic acids are produced as part of bacterial and fungal metabolic reaction schemes[14] and, similar to siderophores, can be used to either increase the availability of metals or chelate them with the purpose of reducing metal toxicity to the organism. However, the zone of effectiveness of the first two types of ligands as mobilizers in soils may be limited to the rhizosphere and biofilms, i.e., areas with an abundance of microorganisms able to produce and consume metal-chelating ligands.[197] The high surface area and generally low nutrient concentrations in soil systems are ideal for biofilm development[199] and thus for production of the characteristic polymeric matrix in both aerobic and anaerobic environments. Aside from providing stability and adherence to the substratum, this matrix provides adsorption sites for organic and inorganic nutrients and chemical species (including metals) by means of carboxyl and hydroxyl functional groups of various types.[198] Through sloughing off surfaces, the exopolymer/metal complexes thus released are often soluble or colloidal. In a study on metal binding characteristics of the capsular exopolymer of *Bacillus licheniformis*, Cu^{2+}, Al^{3+}, Cr^{3+}, and Fe^{3+} showed the greatest affinity for the polymer in competitive-binding experiments.[200] The metals can be released from these complexes via simple diffusion, removal by ligands with greater binding abilities, and microbial degradation of the polysaccharides. In an anaerobic environment, the bound metals will be reduced and subsequently solubilized. Aside from indirect evidence of metal mobilization, direct evidence was obtained by Chanmugathas and Bollag[201] who demonstrated that cadmium was transported in large part by high and low molecular weight organic molecules.

5.2 Mercury, Selenium, and Arsenic

These inorganic compounds have been grouped together, as similar transformation and detoxification mechanisms have been described to occur. In the environment, selenium and arsenic exist both in the inorganic and organic state, and their biogeochemistry is similar to that of sulfur and phosphorus, respectively. The biogeochemistry of mercury, a liquid element at 20°C, stands apart from both previous metals due to its unusual chemical nature. The different valence states of inorganic mercury (Hg°, Hg^{+}, and Hg^{2+}) exist in chemical equilibrium, as described by Robinson and Tuovinen.[186] Sources of organo-mercuric, -selenoic, and -arsenoic compounds can either be of an anthropogenic (e.g., agricultural, pharmaceutical, pulp and paper industries, …) or of a biogenic nature. Microorganisms, including bacteria and fungi, have been shown to either methylate mercury, selenium, and arsenic or incorporate these metals into amino acids and more complex organic compounds.[185,186]

Of all microbial transformations of heavy metals, mercury has been studied most extensively. Although the formation of methylmercury has been demonstrated under both aerobic and anaerobic conditions, it was faster and resulted in higher net levels under anaerobic conditions and in samples with the highest organic carbon content.[186] Moreover, Berman et al.[202] recently demonstrated that more than 95% of mercury methylation was tied to the activity of dissimilatory sulfate reducers. The mechanism of methylation appears to involve the nonenzymatic transfer of methyl groups from microbially excreted methylcobalamin (vitamin B_{12}) to Hg^{2+}. The reaction was shown to require ATP and hydrogen as the source of electrons. Aside from microbial mediation in methylmercury production, transmethylation from other organometals (e.g., organotins) or by humic substances may constitute forms of abiotic methylation mechanisms.[183] In the environment, the availability and presence of methylmercury are regulated by the relative rates of its production (i.e., methylation) and decomposition (i.e., demethylation).

Demethylation of methylmercury is tied to the volatilization of mercury, as metallic mercury (Hg°) is usually found as the volatile end product of mercury detoxification (i.e., the elimination of the metal). Volatilization of organo-mercurial compounds (e.g., methylmercury) is accomplished via a two-step reaction: the organomercurial lyase enzyme cleaves the carbon-mercury linkage (in the case of methylmercury, to methane and Hg^{2+}), followed by the reduction of Hg^{2+} to $Hg°$ by the mercuric reductase enzyme. Both enzymes are inducible (but not by HgS) and are inhibited by Cd^{2+} and Cu^{2+}. Hence, the rate of mercury volatilization from a site contaminated with a complex mixture of heavy metals may be reduced, if not completely inhibited. Recently, a new oxidative demethylation mechanism was reported involving sulfate reducers and methanogenic bacteria isolated from freshwater and estuarine sediments.[203] Where aerobes were the most significant demethylating organisms in estuarine sediments, anaerobic demethylation occurred in freshwater sediments resulting in the evolution of CO_2 and lesser amounts of methane.

Selenium is considered one of the least plentiful, but most toxic elements in the earth's crust.[190] The forms of selenium in soil include metal selenides (e.g., CuSe), elemental selenium ($Se°$), selenite (SeO_3^{2-}), selenate (SeO_4^{2-}), and organic selenium. Although domestic and industrial wastes contribute to selenium release in the environment, one of the most publicized cases of selenium contamination — the Kesterson Reservoir (San Joaquin Valley, California) — derived its ecologically calamitous concentrations from dissolution and accumulation of naturally occurring seliniferous material in agricultural drainage water. Selenium appears to be predominantly cycled via the biological pathways mentioned earlier.[190] Whereas few reports are available on the oxidation of selenium, the microbial reduction of oxidized, inorganic selenium compounds usually results in incorporation of selenide into organic compounds (i.e., immobilization) or the formation of elemental selenium. Under anaerobic conditions, evolution of hydrogen selenide (H_2Se) was observed from soils amended with selenate, selenite, or elemental selenium.[204] Oremland et al.[205,206] recently reported on the rapid in situ selenate reduction in anoxic sediments from agricultural drainage systems exposed to nanomolar concentrations of selenium oxyanions.

Contrary to mercury methylation, which is mainly affected by sulfate-reducing bacteria, soil fungi are the predominant group of microorganisms affecting selenium methylation forming dimethyl selenide and dimethyl diselenide as the volatile end products. Though methylation has been demonstrated under both aerobic and anaerobic conditions, the process is slower under anaerobic conditions and is affected by concurrent reduction processes (to elemental Se or H_2Se gas). Moreover, inorganic selenium forms are less readily transformed to dimethylselenide than are organic compounds. The evolution of volatile selenium (mainly dimethyl selenide) from unamended and selenite-amended soils is directly related to the content of soluble selenium and — in the case of Kesterson Reservoir — was stimulated by the addition of available carbon sources (e.g., gluten, casein, manure), temperature (optimum 28 to 30°C), and acidity (pH 6.5 to 7.0).[207,208] In a study on the volatilization of alkylselenides from soils (derived from a selenium-problem area) spiked with labeled selenium, Karlson and Frankenberger[191] found that the fraction of added selenium volatilized per unit time was dependent upon the soil type, selenium species, and carbon amendment.

Without amendment, Se (IV) was volatilized faster than Se (VI), a difference which could be adjusted by providing pectin as a carbon source (presumably to increase the active microbial population). No minimum threshold for selenium volatilization was found (down to 10 μg Se per kilogram soil).

The biogeochemistry and chemical behavior of arsenic (oxidation states As (V), arsenate, and As (III), arsenite) is analogous to that of phosphorus. Hence, the detoxification mechanisms developed to prevent interference of arsenic with phosphorus-dependent biological processes include the production of organo-arsenicals, and incorporation of arsenic in arsonium compounds (e.g., arsenobetaine, arsenocholine).[188] Organo-arsenicals include mono-, di-, and trimethylated species with methylation apparently occurring in nutrient-depleted environments.[209] Methylation has been shown to occur primarily in aerobic estuarine waters and to a limited extent in freshwaters. Under anoxic, sulfate-reducing conditions such as exist in lake sediments, arsenate can be reduced to arsenite or even removed as an arsenic-enriched mineral phase. Moreover, the evolution of arsine (AsH_3) has been demonstrated under anaerobic conditions.[210] Both oxidation states of arsenic have been shown to be subjected to oxidation/reduction and methylation reactions in soils, both by fungi and presumably sulfate-reducing bacteria. As the highest concentration (up to 50 mg/kg soil) of soluble arsenic species (oxidation states III and V) in soils occurs at a redox potential (E_h) between 0 and 100 mV,[211] methylation (and not reduction) can be anticipated to be the predominant detoxification mechanism in seasonally anoxic soils (Figure 1).

A summary of the physiological conditions and requirements to carry out all described types of detoxification and transformation mechanisms is presented in Table 5.

5.3 Chromium, Cadmium, and Cyanide-Complexes

The principal detoxification processes of these metals are either bacterial reductions from more soluble — and thus more toxic — to less soluble valence states as is the case with chromium (Cr^{6+} to Cr^{3+})[212] or the formation of organic complexes to reduce the bioavailability. Both aerobic[213] and anaerobic[214] chromate reductions have been reported. The reduced trivalent chromium at neutral pH readily forms less-soluble chromium-hydroxides. Whether the anaerobic microorganisms are able to use chromium as a terminal electron acceptor is unknown. Thus far, chromium transformations have only been demonstrated in controlled laboratory incubations with axenic cultures.[215]

Little is known about the fate of cadmium after its introduction into the soil. Short-term exposure will result in the binding of cadmium to soil. However, long-term exposure might result in remobilization of the sorbed metal through the formation of complexes with low and high molecular weight organic ligands, microbial metabolites, or decomposed products. Recently, Chanmugathas and Bollag[216] provided the first direct evidence of heavy metal immobilization and subsequent mobilization by microbial action in acid sandy soil columns. This process was shown to occur both under aerobic and anaerobic conditions; the

Table 5.   Aerobic and Anaerobic Transformations of Heavy Metals

Reaction Type	Examples	Product(s)	Environmental Conditions
Methylations	Mercury	Mono-, dimethyl mercury	Anaerobic conditions, presence of SRB,[a] complicated by the presence of sulfide
	Selenium	Dimethyl selenide	Organic C-amendment, soil fungi
	Arsenic, lead, tin	Mono-, di-, trimethylated	Nutrient depleted environments
Demethylations	Mercury	Divalent Hg	Cd^{2+}, Cu^{2+}/prior exposure to organo-Hg compounds required
	Selenium	Methane, carbon dioxide	Methanogenic conditions
Oxidations	Mercury	Divalent Hg, organic ligands	Oxygen
	Selenium	Selenic acid (H_2SeO_4)	Oxygen
	Arsenic, lead, tin		
	Cyanide, metallo-CN	Ammonia, bicarbonate, $Ni(CN)_2$	Methanogenic conditions
Reductions	Mercury	Volatile $Hg°$	Inhibition by Cd^{2+} and Cu^{2+} prior exposure to organo-Hg compounds required
	Selenium	Insoluble $Se°$, hydrogen selenide	Presence of SRB
	Arsenic	Arsine (AsH_3), insoluble arsenite	
	Chromium	Cr(III)hydroxides, free Cr^{3+}	
Immobilization	Selenium	Methylated products, biological macromolecules	Aerobic/anaerobic conditions
			Competition with S in biosynthesis
	Arsenic	Methylated products, biological macromolecules	Competition with P in biosynthesis
	Cadmium	Complex formation with organic	Anaerobic/aerobic conditions
	Ligands	Aerobic mineralization	

a   SRB: Sulfate reducing bacteria.

former resulted in a decreased pH and concurrent mobilization of cadmium, presumably caused by the aerobic degradation of the organo-metal complex.[201] This could result from a detoxification mechanism by microorganisms to convert cadmium to a form that is either less toxic or less able to enter the cell. Anaerobically, cadmium was immobilized without significant remobilization, while the pH remained unchanged.

A last form of inorganic waste considered in this chapter concerns the physiology of free cyanide and metal-cyano complexes degradation. Coal coking, precious metals mining, and nitrile polymer industries have been recognized as the responsible sources for cyanide input into the environment. Since cyanide is highly reactive, it will readily bind metals as a strong ligand to form metal-cyano complexes (e.g., the tetracyano complexes of divalent nickel, copper, and zinc; the hexacyano complexes of di- and trivalent iron, and related derivatives of chromium).[217] Free cyanide (CN-) has been shown to be transformed and detoxified under both aerobic and anaerobic conditions.[218] Physiologically, cyanide (>10 mg/L) has been found to be inhibitory to methanogenesis resulting in long lag phases, in studies using phenolic compounds as the primary reduced carbon sources.[219,220] However, methanogenesis was maintained in continuous anaerobic columns adapted to >100 mg/L cyanide with ethanol, phenol, or methanol as primary substrates. Ammonia was found as the nitrogen end product, and bicarbonate was found as the major carbon end product of cyanide transformation.[221-222]

Recently, the degradation of a metal-cyano complex has been investigated. Tetracyanonickelate (II) was shown to be used as the sole nitrogen source (substrate range: 0.25 to 16 mM) by microorganisms isolated from soil, freshwater, and sewage sludge enrichments. Overall, the results showed: (1) the degradation appeared to be similar to growth on potassium cyanide (KCN); (2) degradation does not occur in the presence of ammonia as an alternative N-source; and (3) nickel cyanide ($Ni(CN)_2$) accumulated as a major biodegradation product.[217]

5.4 Uranium and Cesium

Surface waters or groundwaters may have undesirable high concentrations of dissolved uranium as the result of natural processes, of uranium mining and processing, or the release of nuclear materials in the environment. Similarly, radioactive cesium has been detected in the environment near a weapons plant test area and in wastewater from an energy-producing plant.[223] Although both elements (or the radioactive isotopes) accumulate in the environment, recent findings have shown that microorganisms are able to lower the overall concentration in aqueous media.

Although U (VI) is soluble in most surface and groundwaters at ambient pH, U (IV) is highly insoluble and precipitates out. Both dissimilatory iron (III)-reducing[224,225] and sulfate-reducing[226] microorganisms have been demonstrated to reduce U (VI) under anaerobic conditions. Thus, uranium can potentially be converted from a soluble to the insoluble U (IV). Enzymatic reduction was shown to be faster than abiotic reduction. Whereas the iron (III)-reducing microorganisms were able to use U (VI) as a terminal electron acceptor (acetate or H_2 as electron donors) and were able to obtain energy and support growth from this metabolism,[225] the sulfate-reducers could not grow at the expense of U (VI) and precipitated extracellular mineral U (VI) uranite.[226] In the latter case, lactate or H_2 was supplied as the electron donors.

Cesium cannot be modified, rather is accumulated by microorganisms. The concentration factors during bioaccumulation of ^{137}Cs have been reported to average around 4.6×10^2. Although the process of bioaccumulation has been described earlier in algae (e.g., *Euglena* sp. and *Chlorella* sp.[227]), cesium-accumulating bacteria have only recently been isolated.[228] Moreover, the concentration factor in the isolated *Rhodococcus* sp. was at least one order of magnitude higher than the values reported for algae. Bioaccumulation processes could, therefore, be useful for the removal and monitoring of radioactive elements such as neptunium, uranium, and cesium.

6. CONCLUSIONS

It is apparent from this discussion that the application of biotechnology for the biological destruction of recalcitrant, synthetic, and at times toxic components associated with environmentally hazardous wastes is not immediately transferable from our knowledge of conventional waste treatment process engineering, which deals with "easily metabolizable" or labile substrates. The influence of a wide range of biological and environmental parameters on the biodegradative potential of a given microbial population has been reviewed and assessed for possible stimulation in general and for specific groups of target contaminants in particular.

The decision to consider the option of bioremediation for a given type of contamination still has to rely largely on information obtained from laboratory studies, with the exception perhaps of petroleum hydrocarbons. Scattered field scale studies such as for TCE, BTEX compounds, and PCBs are starting to emerge and provide site-specific information on the types and capabilities of microbial populations present, the acclimation period required before degradation commences, and the extent of degradation that can be expected. Information is needed on the level and distribution of the pollutants, on the presence of more "labile" — thus competing — carbon sources, and on those environmental parameters influencing aerobic, anaerobic, and microaerophilic conditions. Moreover, our knowledge of monitoring the activity of *in situ* stimulated microorganisms is still scant. Aside from the appearance of metabolites, which indirectly indicates microbial activity, actual proof that the microorganisms are the causative agents for substrate disappearance or metabolite appearance is difficult to obtain. Recently, Madsen et al.[229] described a method to prove that the mass of pollutant compounds (in this case PAH contamination) was degraded by microorganisms. The authors analyzed a comprehensive set of microbiological activity and distribution data and concluded that enhanced numbers of protozoa and their bacterial prey were found exclusively in the contaminated subsurface samples. The main drawback to this method is that it would be applicable only in those cases where the contaminant(s) serve(s) as (a) growth substrate(s) for the microorganisms, thus excluding cometabolic degradation processes. If no indigenous microbial population is able to effect the necessary transformation steps of the pollutants, a laboratory-cultivated (previously isolated) culture may be required to inoculate the contaminated site. For those microorganisms, the same physiological requirements described above apply to ensure degradative activities.

A reality that needs to be addressed is that biologically mediated transformations often lead to intermediates, not necessarily listed as contaminants, which may be equally hazardous or more mobile in the soil environment. Although the aim of bioremediation is mineralization to CO_2, this is often not the case, especially where secondary metabolism and cometabolism of the pollutants are concerned, as these metabolic processes are not beneficial to the microbial populations present. Thus, biological degradation cannot always be expected to reach the same degree of removal efficiency and destruction when compared to other remediation technologies. Whether this reality is an allowable sacrifice for the lower economic cost of environmental biotechnology has to be carefully considered in light of the specifics of the site and type of contamination concerned.

7. REFERENCES

1. Stotzky, G. "Activity, Ecology, and Population Dynamics of Microorganisms in Soil," in *Microbial Ecology*, A. Laskin and H. Lechevalier, Eds. (Cleveland, OH: CRC Press, 1974), pp. 57–135.
2. Grbić-Galić, D. "Anaerobic Microbial Transformation of Nonoxygenated Aromatic and Alicyclic Compounds in Soil, Subsurface, and Freshwater Sediments," in *Soil Biochemistry, Volume 6*, J.-M. Bollag and G. Stotzky, Eds. (New York: Marcel Dekker, Inc., 1990), pp. 117–189.
3. Boethling, R.S., and M. Alexander. "Effect of Concentration of Organic Chemicals on their Biodegradation by Natural Microbial Communities," *Appl. Environ. Microbiol.* 37:1211–1216 (1979).
4. Alexander, M. "Biodegradation of Chemicals of Environmental Concern," *Science* 211:132–138 (1981).
5. Wang, Y., R. V. Subba-Rao, and M. Alexander. "Effect of Substrate Concentration and Organic and Inorganic Compounds on the Occurrence and Rate of Mineralization and Cometabolism," *Appl. Environ. Microbiol.* 47:1195–1200 (1984).
6. McCarty, P. L. "Bioengineering Issues Related to In-Situ Remediation of Contaminated Soils and Groundwater," in *Environmental Biotechnology: Reducing Risks from Environmental Chemicals through Biotechnology* , G. S. Omenn, Ed. (New York: Plenum Press, 1988), pp. 143–163.
7. Swindoll, C. M., C. M. Aelion, and F. K. Pfaender. "Influence of Inorganic and Organic Nutrients on Aerobic Biodegradation and on the Adaptation Response of Subsurface Microbial Communities," *Appl. Environ. Microbiol.* 54:212–217 (1988).

8. Turco, R. F., and A. Konopka. "Biodegradation of Organic Compounds in the Vadose Zone and Aquifer Sediments," *Appl. Environ. Microbiol.* 57:2260–2268 (1991).

9. Wilson, J. T., J. F. McNabb, D. L. Balkwill, and W. C. Ghiorse. "Enumeration and Characterization of Bacteria Indigenous to Shallow Water Table Aquifers," *Groundwater* 21:134–142 (1983).

10. Federle, T. W., D. C. Dobbins, J. R. Thornton-Manning, and D. D. Jones. "Microbial Biomass, Activity, and Community Structure in Subsurface Soils," *Groundwater* 24:365–374 (1986).

11. Scopp, J., M. D. Jawson, and J. W. Doran. "Steady-State Aerobic Microbial Activity as a Function of Soil Water Content," *Soil Sci. Soc. Am. J.* 54:1619–1625 (1990).

12. Brock, T. D., D. W. Smith, and M. T. Madigan. *Biology of Microorganisms* (Englewood Cliffs, NJ: Prentice Hall, Inc., 1984), p. 847.

13. Schlegel, H. G. *Allgemeine Microbiologie* (Stuttgart, Germany: Thieme Verlag, 1985), p. 569.

14. Gottschalk, G. *Bacterial Metabolism* (New York: Springer Verlag, 1986), p. 359.

15. Alexander, M. "Biodegradation: Problems of Molecular Recalcitrance and Microbial Fallibility," *Adv. Appl. Microbiol.* 7:35–80 (1965).

16. Horvath, R. S. "Microbial Cometabolism and the Degradation of Organic Compounds in Nature," *Bacteriol. Rev.* 36:146–155 (1972).

17. Shelton, D. R., and J. M. Tiedje. "General Method for Determining Anaerobic Biodegradation Potential," *Appl. Environ. Microbiol.* 47:850–857 (1984).

18. Zehnder, A. J. B., and W. Stumm. "Geochemistry and Biogeochemistry of Anaerobic Habitats," in *Biology of Anaerobic Microorganisms,* A. J. B. Zehnder, Ed. (New York: John Wiley and Sons, 1988), pp. 1–38.

19. Oremland, R. S., and B. F. Tayler. "Sulfate Reduction and Methanogenesis in Marine Sediments," *Geochim. Cosmochim. Acta* 42:209–214 (1978).

20. Barnes, R. O., and E. D. Goldberg. "Methane Production and Consumption in Anoxic Sediments," *Geology* 4:297–300 (1976).

21. Martens, C. S., and R. A. Berner. "Methane Production in the Interstitial Waters of Sulfate-Depleted Marine Sediments," *Science* 185:1167–1169 (1974).

22. Winfrey, J. R., and J. G. Zeikus. "Effect of Sulfate on Carbon and Electron Flow during Microbial Methanogenesis in Freshwater Sediments," *Appl. Environ. Microbiol.* 33:275–281 (1977).

23. Widdel, F. "Microbiology and Ecology of Sulfate and Sulfur Reducing Bacteria," in *Biology of Anaerobic Microorganisms,* A. J. B. Zehnder, Ed. (New York: John Wiley and Sons, 1988), pp. 469–578.

24. Oremland, R. S. "Biogeochemistry of Methanogenic Bacteria," in *Biology of Anaerobic Microorganisms,* A. J. B. Zehnder, Ed. (New York: John Wiley and Sons, 1988), pp. 641–707.

25. Young. L. Y. "Anaerobic Degradation of Aromatic Compounds," in *Microbial Degradation of Organic Compounds,* D. T. Gibson, Ed. (New York: Marcel Dekker, Inc., 1984), pp. 487–523.

26. Schink, B. "Principles and Limits of Anaerobic Degradation: Environmental and Technological Aspects," in *Biology of Anaerobic Microorganisms,* A. J. B. Zehnder, Ed. (New York: John Wiley and Sons, 1988), pp. 771–841.

27. Vogel, T. M., C. S. Criddle, and P. L. McCarty. "Transformation of Halogenated Aliphatic Compounds," *Environ. Sci. Technol.* 8:722–736 (1987).

28. Hill, D. W. and P. L. McCarty. "Anaerobic Degradation of Selected Chlorinated Hydrocarbon Pesticides," *J. Wat. Pollut. Cont. Fed.* 39:1259–1277 (1967).

29. Suflita, J. M., A. Horowitz, D. R. Shelton, and J. M. Tiedje. "Dehalogenation: A Novel Pathway for the Anaerobic Biodegradation of Haloaromatic Compounds," *Science* 218:1115–1117 (1982).

30. Gantzer, C. J., and L. P. Wackett. "Reductive Dechlorination Catalyzed by Bacterial Transition-Metal Coenzymes," *Environ. Sci. Technol.* 25:715–722 (1982).

31. Assaf-Anid, N., L. Nies, and T. M. Vogel. "Reductive Dechlorination of a Polychlorinated Biphenyl Congener and Hexachlorobenzene by Vitamin B_{12}," *Appl. Environ. Microbiol.* 58:1057–1060 (1992).

32. Castro, C. E. "Biodehalogenation," *Environ. Health Perspect.* 21:279–283 (1977).

33. Criddle, C. S. "Reductive Dehalogenation in Microbial and Electrolytic Systems," Ph.D. Thesis, Stanford University, Stanford, CA (1989).

34. Rochkind-Dubinsky, M. L., G. S. Sayler, and J. W. Blackburn. *Microbial Decomposition of Chlorinated Aromatic Compounds* (New York: Marcel Dekker Inc., 1987), p. 367.

35. Baxter, R. M. "Reductive Dehalogenation of Environmental Contaminants: A Critical Review," *Water Pollut. Res. J. Canada* 24:299–322 (1989).

36. Bryant, F. O., D. D. Hale, and J. E. Rogers. "Regiospecific Dechlorination of Pentachlorophenol by Dichlorophenol-Adapted Microorganisms in Freshwater, Anaerobic Sediment Slurries," *Appl. Environ. Microbiol.* 57:2293–2301 (1991).

37. Dolfing, J. "Reductive Dechlorination of 3-Chlorobenzoate Is Coupled to ATP Production and Growth in an Anaerobic Bacterium, Strain DCB-1," *Arch. Microbiol.* 153:264–266 (1990).

38. Mohn, W. W., and J. M. Tiedje. "Strain DCB-1 Conserves Energy for Growth from Reductive Dechlorination Coupled to Formate Oxidation," *Arch. Microbiol.* 153:267–271 (1990).

39. Atlas, R. M. "Microbial Degradation of Microbial Hydrocarbons: An Environmental Perspective," *Microbiol. Rev.* 45:180–209 (1981).

40. Miller, R. M., and R. M. Bartha. "Evidence for Liposome Encapsulation for Transport-Limited Microbial Metabolism of Solid Alkanes," *Appl. Environ. Microbiol.* 55:269–274 (1989).

41. Hommel, R. K. "Formation and Physiological Role of Biosurfactants Produced by Hydrocarbon-Utilizing Microorganisms," *Biodegradation* 1:107–119 (1990).

42. Watkinson, R. J., and P. Morgan. "Physiology of Aliphatic Hydrocarbon-Degrading Microorganisms" *Biodegradation* 1:79–93 (1990).

43. Finnerty, W. R., and M. E. Singer. "Membranes of Hydrocarbon Utilizing Microorganisms," in *Organization of Prokaryotic Cell Membranes, Volume III*, B. K. Ghosh, Ed. (Boca Raton, FL: CRC Press, Inc., 1985), pp. 1–44.

44. Hommel, R., O. Stuewer, W. Stuber, D. Haferburg, and H.-P. Kleber. "Production of Water-Soluble, Surface-Active Exolipids by *Torulopsis apicola*," *Appl. Microb. Biotechnol.* 26:199–205 (1987).

45. Itoh, S., H. Honda, F. Tomita, and T. Suzuki. "Rhamnolipid Produced by *Pseudomonas aeruginosa* Grown on n-Paraffin," *J. Antibiot.* 24:855–859 (1971).

46. Itoh, S., and T. Suzuki. "Effects of Rhamnolipids on Growth of *Pseudomonas aeruginosa* Mutant Deficient in n-Paraffin-Utilizing Ability," *Agric. Biol. Chem.* 38:1443–1449 (1972).

47. Itoh, S., and S. Inoue. "Sophorolipids from *Torulopsis bombicola*: Possible Relation to Alkane Uptake," *Appl. Environ. Microbiol.* 43:1278–1283 (1982).

48. Fuhs, G. W. "Der Mikobielle Abbau von Kohlenwasserstoffen," *Arch. Microbiol.* 39:374–422 (1961).

49. Thauer, R. K., K. Jungermann, and K. Decker. "Energy Conservation in Chemotrophic Anaerobic Bacteria," *Bacteriol. Rev.* 41:100–180 (1977).

50. Zehnder, A. J. B., and T. D. Brock. "Methane Formation and Methane Oxidation by Methanogenic Bacteria," *J. Bacteriol.* 137:420–432 (1979).

51. Gibson, D. T. "Microbial Degradation of Hydrocarbons," in *The Nature of Seawater (Dahlem Konferenzen, Berlin)*, E. D. Goldberg, Ed. (Heidelberg, Germany: Springer Verlag, 1975), pp. 140–155.

52. Perry, J. J. "Microbial Cooxidations Involving Hydrocarbons," *Microbiol. Rev.* 43:59–79 (1979).

53. Hansen, R. W., and R. E. Kallio. "Inability of Nitrate to Serve as a Terminal Oxidant for Hydrocarbons," *Science* 125:1198–1199 (1957).

54. Culbertson, C. W., A. J. B. Zehnder, and R. S. Oremland. "Anaerobic Oxidation of Acetylene by Estuarine Sediments and Enrichment Cultures," *Appl. Environ. Microbiol.* 41:396–403 (1981).

55. Schink, B. "Degradation of Unsaturated Hydrocarbons by Methanogenic Enrichment Cultures," *FEMS Microbiol. Ecol.* 31:69–77 (1985).

56. Britton, L. N. "Microbial Degradation of Aliphatic Hydrocarbons," in *Microbial Degradation of Organic Compounds*, D. T. Gibson, Ed. (New York: Marcel Dekker, Inc., 1984), pp. 89–131.

57. Gibson, D. T., and V. Subramanian. "Microbial Degradation of Aromatic Hydrocarbons," in *Microbial Degradation of Organic Compounds*, D. T. Gibson, Ed. (New York: Marcel Dekker, Inc., 1984), pp. 181–253.

58. Smith, M. R. "The Biodegradation of Aromatic Hydrocarbons by Bacteria," *Biodegradation* 1:191–206 (1990).

59. Suflita, J. M., and G. W. Sewell. "Anaerobic Biotransformation of Contaminants in the Subsurface," U.S. EPA Environmental Research Brief 600/M-90/024 (1990).

60. Colberg, P. J. "Anaerobic Microbial Degradation of Cellulose, Lignin, Oligolignols, and Monoaromatic Lignin Derivatives," in *Biology of Anaerobic Microorganisms*, A. J. B. Zehnder, Ed. (New York: John Wiley and Sons, 1988), pp. 469–578.

61. Guyer, M., and G. Hegeman. "Evidence for a Reductive Pathway for the Anaerobic Metabolism of Benzoate," *J. Bacteriol.* 99:906–907 (1969).

62. Williams, R. J., and W. C. Evans. "The Metabolism of Benzoate by *Moraxella* sp. through Anaerobic Nitrate Respiration: Evidence for a Reductive Pathway," *Biochem. J.* 148:1–10 (1975).

63. Widdel, F. "Anaerober Abbau von Fettsauren und Benzoesaure durch neu Isolierten Arten Sulfat-Reduzierender Bacterien," Ph.D. Thesis, University of Goettingen, Germany (1980).

64. Dean, B. J. "Recent Findings on the Genetic Toxicology of Benzene, Toluene, Xylenes and Phenols," *Mutat. Res.* 154:153–181 (1985).

65. Ward, D. M., R. M. Atlas, P. D. Boehm, and J. A. Calder. "Biodegradation and Chemical Evolution of Oil from the Amoco Microbial Spill," *AMBIO, J. Human Environ. Res. Manage.* 9:277–283 (1980).

66. Kuhn, E. P., P. J. Colberg, J. L. Schnoor, O. Wanner, A. J. B. Zehnder, and R. P. Schwartzenbach. "Microbial Transformations of Substituted Benzenes during Infiltration of River Water to Ground Water: Laboratory Column Studies," *Environ. Sci. Technol.* 19:961–968 (1985).

67. Zeyer, J., E. P. Kuhn, and R. P. Schwartzenbach. "Rapid Microbial Mineralization of Toluene and 1,3-Dimethylbenzene in the Absence of Molecular Oxygen," *Appl. Environ. Microbiol.* 52:944–947 (1986).

68. Dolfing, J., J. Zeyer, P. Binder-Eicher, and R. P. Schwartzenbach. "Isolation and Characterization of a Bacterium that Mineralizes Toluene in the Absence of Molecular Oxygen," *Arch. Microbiol.* 154:336–341 (1990).

69. Evans, P. J., D. T. Mang, K. S. Kim, and L. Y. Young. "Anaerobic Degradation of Toluene by a Denitrifying Bacterium," *Appl. Environ. Microbiol.* 57:1139–1145 (1991).

70. Evans, P. J., D. T. Mang, and L. Y. Young. "Degradation of Toluene and *m*-Xylene and Transformation of *o*-Xylene by Denitrifying Enrichment Cultures," *Appl. Environ. Microbiol.* 57:450–454 (1991).

71. Hutchins, S. R., G. W. Sewell, D. A. Kovacs, and G. A. Smith. "Biodegradation of Aromatic Hydrocarbons by Aquifer Microorganisms under Denitrifying Conditions," *Environ. Sci. Technol.* 25:68–76 (1991).

72. Ramanand, K., and J. M. Suflita. "Anaerobic Degradation of *m*-Cresol in Anoxic Aquifer Slurries: Carboxylation Reactions in a Sulfate-Reducing Bacterial Enrichment," *Appl. Environ. Microbiol.* 57:1689–1695 (1991).

73. Haag, F., M. Reinhard, and P. L. McCarty. "Degradation of Toluene and *p*-Xylene in Anaerobic Microcosms: Evidence for Sulfate as a Terminal Electron Acceptor," *Environ. Toxicol. Chem.* 10:1379–1389 (1991).

74. Beller, H. R., D. Grbić-Galić, and M. Reinhard. "Microbial Degradation of Toluene under Sulfate-Reducing Conditions and the Influence of Iron on the Process," *Appl. Environ. Microbiol.* 58:786–793 (1992).

75. Edwards, E. A., L. E. Wills, M. Reinhard, and D. Grbić-Galić. "Anaerobic Degradation of Toluene and Xylene by Aquifer Microorganisms under Sulfate-Reducing Conditions," *Appl. Environ. Microbiol.* 58:794–800 (1992).

76. Lovley, D. R., M. J. Baedecker, D. J. Lonergan, I. M. Cozzarelli, E. J. P. Phillips, and D. I. Siegel. "Oxidation of Aromatic Contaminants Coupled to Microbial Iron Reduction," *Nature (London)* 339:297–300 (1989).

77. Lovley, D. R., and D. J. Lonergan. "Anaerobic Oxidation of Toluene, Phenol, and *p*-Cresol by the Dissimilatory Iron-Reducing Organism, GS-15," *Appl. Environ. Microbiol.* 54:490–496 (1990).

78. Wilson, B. H., G. B. Smith, and J. F. Rees. "Biotransformation of Selected Alkylbenzenes and Halogenated Aliphatic Hydrocarbons in Methanogenic Aquifer Material: A Microcosm Study," *Environ. Sci. Technol.* 20:997–1002 (1986).

79. Vogel, T. M., and D. Grbić-Galić. "Incorporation of Oxygen from Water into Toluene and Benzene during Anaerobic Fermentative Transformation," *Appl. Environ. Microbiol.* 52:200–202 (1986).

80. Grbić-Galić, D., and T. M. Vogel. "Transformation of Toluene and Benzene by Mixed Methanogenic Cultures," *Appl. Environ. Microbiol.* 53:254–260 (1987).

81. Ball, H. A., M. Reinhard, and P. L. McCarty. "Biotransformation of Monoaromatic Hydrocarbons under Anoxic Conditions," in *In Situ Bioreclamation*, R. E. Hinchee and R. F. Offenbuttle, Eds. (Boston: Butterworth-Heinemann, Inc., 1991), pp. 458–463.

82. Major, D. W., C. I. Mayfield, and J. F. Barker. "Biotransformation of Benzene by Denitrification in Aquifer Sand," *Ground Water* 26:8–14 (1988).

83. Roberts, D. J., P. M. Fedorak, and S. E. Hrudey. "CO_2 Incorporation and 4-Hydroxy-2-Methylbenzoic Acid Formation during Anaerobic Metabolism of *m*-Cresol by a Methanogenic Consortium," *Appl. Environ. Microbiol.* 56:472–478 (1990).

84. Bisaillon, J.-G., F. Lepine, R. Baudelet, and M. Sylvestre. "Carboxylation of *o*-Cresol by an Anaerobic Consortium under Methanogenic Conditions," *Appl. Environ. Microbiol.* 57:2131–2134 (1991).

85. Evans, P. J., W. Ling, B. Goldsmith, E. R. Ritter, and L. Y. Young. "Metabolites Formed during Anaerobic Transformation of Toluene and their Proposed Relationship to the Initial Steps of Toluene Mineralization," *Appl. Environ. Microbiol.* 58:496–501 (1992).

86. Beller, H. R., M. Reinhard, and D. Grbić-Galić. "Metabolic Byproducts of Anaerobic Toluene Degradation by Sulfate-Reducing Enrichment Cultures," *Appl. Environ. Microbiol.* 58:3192–3195 (1992).

87. Alvarez, P. J. J., and T. M. Vogel. "Substrate Interactions of Benzene, Toluene, and *p*-Xylene during Microbial Degradation by Pure Cultures and Mixed Culture Aquifer Slurries," *Appl. Environ. Microbiol.* 57:2981–2985 (1991).

88. Oldenhuis, R., L. Kuijk, A. Lammers, D. B. Janssen, and B. Witholt. "Degradation of Chlorinated and Nonchlorinated Aromatic Solvents in Soil Suspensions by Pure Bacterial Cultures," *Appl. Micobiol. Biotechnol.* 30:211–217 (1989).

89. Thomas, J. M., V. R. Gordy, S. Fiorenza, and C. H. Ward. "Biodegradation of BTEX in Subsurface Materials Contaminated with Gasoline: Granger, Indiana," *Water Sci. Technol.* 6:53–62 (1990).

90. Baggi, G., P. Barbieri, E. Galli, and S. Tollari. "Isolation of a *Pseudomonas stutzeri* Strain that Degrades *o*-Xylene," *Appl. Environ. Microbiol.* 55:2129–2132 (1987).

91. Healy, J. B., Jr., and L. Y. Young. "Anaerobic Biodegradation of Eleven Aromatic Compounds to Methane," *Appl. Environ. Microbiol.* 38:84–89 (1979).

92. Mihelcic, J. R., and R. G. Luthy. "Degradation of Polycyclic Aromatic Hydrocarbon Compounds under Various Redox Conditions in Soil-Water Systems," *Appl. Environ. Microbiol.* 54:1182–1187 (1988).

93. Mihelcic, J. R., and R. G. Luthy. "Microbial Degradation of Acenaphtene and Naphthalene under Denitrification Conditions in Soil-Water Systems," *Appl. Environ. Microbiol.* 54:1188–1198 (1988).

94. Godsy, E. M., D. F. Goerlitz, and D. Grbić-Galić. "Transport and Degradation of Watersoluble Creosote-Derived Compounds," in *Intermedia Pollutant Transport: Modeling and Field Experiments*, D. T. Allen, Y. Cohen, and I. R. Kaplan , Eds. (New York: Plenum Press, Inc., 1989), pp. 213–236.

95. Black, J. P., and D. Grbić-Galić. "Degradation of Polycyclic Aromatic Hydrocarbons by Mixed Methanogenic Cultures," paper presented at the 1991 meeting of the American Society for Microbiology, Dallas, TX, May 1991.

96. Suflita, J. M., L. Liang, and L. Shi. "The Anaerobic Metabolism of 2-Hydroxybiphenyl by Sulfate-Reducing Bacterial Enrichments," *Curr. Microbiol.* 22:69–72 (1990).

97. Newkome, G. R., and W. W. Pandler. *Contemporary Heterocyclic Chemistry, Synthesis, Reactions and Applications* (New York: Wiley and Sons, Inc., 1982), p. 785.

98. March, J. *Advanced Organic Chemistry* (New York: Wiley and Sons, Inc., 1985).

99. Grbić-Galić, D. "Microbial Degradation of Homocyclic and Heterocyclic Aromatic Hydrocarbons under Anaerobic Conditions," *Dev. Ind. Microbiol.* 30:237–253 (1989).

100. Aislabie, J., A. K. Bey, H. Hurst, S. Rothenburger, and R. M. Atlas. "Microbial Degradation of Quinoline and Methylquinolines," *Appl. Environ. Microbiol.* 56:345–351 (1990).

101. Pereira, W. E., C. E. Rostad, T. J. Leiker, D. M. Updegraff, and J. L. Bennett. "Microbial Hydroxylation of Quinoline in Contaminated Groundwater: Evidence for Incorporation of the Oxygen Atom of Water," *Appl. Environ. Microbiol.* 54:827–829 (1988).

102. Cook, A. M., and R. Huetter. "Degradation of s-Triazines: a Critical View of Biodegradation," in *Degradation of Xenobiotics and Recalcitrant Compounds*, T. Leisinger, A. M. Cook, J. Nuesch, and R. Huetter, Eds. (London: Academic Press, 1981), pp. 237–249.

103. Madsen, E. L., A. J. Francis, and J.-M. Bollag. "Environmental Factors Affecting Indole Metabolism under Anaerobic Conditions," *Appl. Environ. Microbiol.* 54:74–78 (1988).

104. Bak, F., and F. Widdel. "Anaerobic Degradation of Indolic Compounds by Sulfate-Reducing Enrichment Cultures, and Description of *Desulfobacterium indolicum* gen. nov., sp. nov," *Arch. Microbiol.* 146:170–176 (1986).

105. Knezovich, J. P., D. J. Bischop, T. J. Kulp, D. Grbić-Galić, and J. Dewitt. "Anaerobic Microbial Degradation of Acridine and the Application of Remote Fiber Spectroscopy to Monitor the Transformation Process," *Environ. Toxicol. Chem.* 9:1235–1243 (1990).

106. Maka, A., V. L McKinley, J. R. Conrad, and K. F. Fannin. "Degradation of Benzothiophene and Dibenzothiophene under Anaerobic Conditions by Mixed Cultures," paper presented at the 1987 Annual Meeting of the American Society for Microbiology, Atlanta, GA, May 1987.

107. Godsy, E.M., and D. Grbić-Galić. "Biodegradation Pathways of Benzothiophene in Methanogenic Microcosms," in U.S. Geological Survey Toxic Substances Hydrology Program-Proceedings of the Technical Meeting, Phoenix, AZ, (1989), pp. 559–564.

108. Focht, D. D., and M. Alexander. "DDT Metabolites and Analogs: Ring Fission by *Hydrogenomonas* sp.," *Science* 170:91–92 (1970).

109. Focht, D. D., and M. Alexander. "Aerobic Cometabolism of DDT Analogues by *Hydrogenomonas* sp.," *J. Agric. Food. Chem.* 19:20–22 (1971).

110. Wedemeyer, G. "Dechlorination of DDT by *Aerobacter aerogenes*," *Science* 152:647 (1966).

111. Wedemeyer, G. "Dechlorination of 1,1,1-Trichloro-2,2-Bis(p-Chlorophenyl)-Ethane by *Aerobacter aerogenes*. I. Metabolic Products," *Appl. Microbiol.* 15:569–574 (1967).

112. Castro, C. E. "The Rapid Oxidation of Iron (II) Porphyrins by Alkyl Halides. A Possible Mode of Intoxication of Organisms by Alkyl Halides," *J. Am. Chem. Soc.* 86:2310–2311 (1964).

113. Westrick, J. J., J. W. Mello, and R. F. Thomas. "The Groundwater Supply Survey," *J. Am. Wat. Works. Assoc.* 76:52–59 (1984).

114. Khöler, A., A. Jager, H. Willershausen, and H. Graf. "Extracellular Ligninase Has No Role in the Degradation of DDT," *Appl. Microbiol. Biotechnol.* 29:618–620 (1988).

115. Castro, T. F., and T. Yoshida. "Degradation of Organochlorine Insecticides in Flooded Soils in the Philippines," *J. Agric. Food Chem.* 19:1168–1170 (1971).

116. Tsukano, Y., and A. Kobayashi. "Formation of g-BTC in Flooded Rice Field Soils Treated with g-BHC," *Agric. Biol. Chem.* 36:166–167 (1972).

117. Zoro, J. A., J. M. Hunter, G. Eglington, and G. C. Ware. "Degradation of p,p'-DDT in Reducing Environments," *Nature (London)* 247:235–236 (1974).

118. Marks, T. S., J. D. Alpress, and A. Maule. "Dehalogenation of Lindane by a Variety of Porphyrins and Corrins," *Appl. Environ. Microbiol.* 55:1258–1261 (1989).

119. Holmstead, R. L. "Studies of the Degradation of Mirex with an Iron (II) Porphyrin Model System," *J. Agric. Food Chem.* 26:590–595 (1976).

120. Saleh, M. A., M. Brunner, and S. M. Schoberth. "Polychlorobornane Components of Toxaphene: Structure-Toxicity Relations and Metabolic Reductive Dechlorination," *Science* 198:1256–1258 (1986).

121. Castro, C. E., and N. O. Belser. "Biodehalogenation. Reductive Dehalogenation of the Biocides Ethylene Dibromide, 1,2-Dibromo-3-Chloropropane, and 2,3-Dibromobutane in Soil," *Environ. Sci. Technol.* 2:779–783 (1968).

122. Bouwer, E. J., and P. L. McCarty. "Transformation of 1- and 2-Carbon Halogenated Aliphatic Organic Compounds under Methanogenic Conditions," *Appl. Environ. Microbiol.* 45:1286–1294 (1983).

123. Brunner, W., D. Staub, and T. Leisinger. "Bacterial Degradation of Dichloromethane," *Appl. Environ. Microbiol.* 40:950–958 (1980).

124. Stuecki, G., R. Gaelli, H.-R. Ebersold, and T. Leisinger. "Dehalogenation of Dichloromethane by Cell Extracts of *Hyphomicrobium DM2*," *Arch. Microbiol.* 130:366–371 (1981).

125. Kohler-Staub, D., S. Hartmans, R. Gaelli, F. Suter, and T. Leisinger. "Evidence for Identical Dichloromethane Dehalogenases in Different Methylotrophic Bacteria," *J. Gen. Microbiol.* 132:2837–2843 (1986).

126. Bouwer, E. J., and P. L. McCarty. "Transformations of Halogenated Organic Compounds under Denitrification Conditions," *Appl. Environ. Microbiol.* 45:1295–1299 (1983).

127. Yokota, T., H. Fuse, T. Omori, and Y. Minoda. "Microbial Dehalogenation of Haloalkanes Mediated by Oxygenase or Halidohydrolase," *Agric. Biol. Chem.* 50:453–460 (1986).

128. Hardman, D. J., and J. H. Slater. "Dehalogenases in Soil Bacteria," *J. Gen. Microbiol.* 123:117–128 (1981).

129. Semprini, L., G. D. Hopkins, D. B. Janssen, M. Lang, P. V. Roberts, and P. L. McCarty. "In Situ Biotransformation of Carbon Tetrachloride under Anoxic Conditions," U.S. EPA Publication 600/S2–90–060 (1991).

130. Vogel, T. M., and P. L. McCarty. "Biotransformation of Tetrachloroethylene to Trichloroethylene, Dichloroethylene, Vinyl Chloride, and Carbon Dioxide under Methanogenic Conditions," *Appl. Environ. Microbiol.* 49:1080–1083 (1985).

131. Higgins, I. J., D. J. Best, R. C. Hammond, and D. Scott. "Methane-Oxidizing Microorganisms," *Microbiol. Rev.* 45:556–590 (1981).

132. Wilson, J. T., and B. H. Wilson. "Biotransformation of Trichloroethylene in Soil," *Appl. Environ. Microbiol.* 49:242–243 (1985).

133. Higgins, I. J., D. J. Best, and R. C. Hammond. "New Findings in Methane-Utilizing Bacteria Highlight their Importance in the Biosphere and their Commercial Potential," *Nature (London)* 286:561–564 (1980).

134. Fox, B. J., and J. D. Libscomb. "Methane Monooxygenase: A Novel Biological Catalyst for Hydrocarbon Oxidation," in *Biological Oxidation Systems, Vol. 1*, G. Hamilton, C. Reddy, and K. M. Madyastha, Eds. (Orlando, FL: Academic Press, 1990), pp. 367–388.

135. Adriaens, P. and D. Grbić-Galić. "Cometabolic Transformation of Chlorinated Biphenyls and Hydroxybiphenyls by Methanotrophic Groundwater Isolates ," *Biodegradation* in review (1993).

136. Semprini, L., P. V. Roberts, G. D. Hopkins, and P. L. McCarty. "A Field Evaluation of In-Situ Biodegradation of Chlorinated Ethenes. 2. Results of Biostimulation and Biotransformation Experiments," *Ground Water* 28:715–727 (1990).

137. Semprini, L., and P. L. McCarty. "Comparison Between Model Simulations and Field Results for In-Situ Biorestoration of Chlorinated Aliphatics. I. Biostimulation of Methanotrophic Bacteria," *Ground Water* 29:365–374 (1991).

138. Semprini, L., G. D. Hopkins, P. V. Roberts, D. Grbić-Galić, and P. L. McCarty. "A Field Evaluation of In-Situ Biodegradation of Chlorinated Ethenes. III. Studies of Competitive Inhibition," *Ground Water* 29:239–250 (1990).

139. Alvarez-Cohen, L., and P. L. McCarty. "Product Toxicity and Competitive Inhibition Modeling of Chloroform and Trichloroethylene Transformation by Methanogenic Resting Cells," *Appl. Environ. Microbiol.* 57:1031–1037 (1991).

140. Henry, S. M., and D. Grbić-Galić. "Inhibition of Trichloroethylene Oxidation by the Transformation Intermediate Carbon Monoxide," *Appl. Environ. Microbiol.* 57:1770–1776 (1991).

141. Henry, S. M., and D. Grbić-Galić. "Influence of Endogenous and Exogenous Electron Donors and Trichloroethylene Oxidation Toxicity on Trichloroethylene Oxidation by Methanotrophic Cultures from a Groundwater Aquifer," *Appl. Environ. Microbiol.* 57:236–244 (1991).

142. Roberts, P. V., G. D. Hopkins, D. M. Mackay, and L. Semprini. "A Field Evaluation of In-Situ Biodegradation of Chlorinated Ethenes. II. Results of Biostimulation and Biotransformation Experiments," *Ground Water* 28:591–604 (1990).

143. McCarty, P. L., L. Semprini, M. Dolan, T. C. Harmon, C. Tiedemen, and S. M. Gorelick. "In Situ Methanotrophic Bioremediation for Contaminated Groundwater at St. Joseph, Michigan," in *On Site Bioreclamation: Processes for Xenobiotic and Hydrocarbon Treatment*, R. E. Hinchee and R. F. Offenbuttle, Eds. (Boston: Butterworth-Heinemann, 1991).

144. Neilson, A. H. "The Biodegradation of Halogenated Organic Compounds," *J. Appl. Bacteriol.* 69:445–470 (1990).

145. Commandeur, L. C. M., and J. R. Parsons. "Degradation of Halogenated Aromatic Compounds," *Biodegradation* 1:207–220 (1990).

146. Bollag, J.-M., and M. J. Loll. "Incorporation of Xenobiotics in Soil Humus," *Experientia* 39:1221–1231 (1983).

147. Focht, D. D., and W. Brunner. "Kinetics of Biphenyl and Polychlorinated Biphenyl Metabolism in Soil," *Appl. Environ. Microbiol.* 50:1058–1063 (1985).

148. Stott, D. E., J. P. Martin, D. D. Focht, and K. Haider. "Biodegradation, Stabilization in Humus, and Incorporation into Soil Biomass of 2,4-D and Chlorocatechol Carbons," *Soil Sci. Soc. Am. J.* 47:66–70 (1983).

149. Furukawa, K. "Microbial Degradation of Polychlorinated Biphenyls," in *Biodegradation and Detoxification of Environmental Pollutants*, A. M. Chakrabarty, Ed. (Boca Raton, FL: CRC Press, 1982), pp. 33–59.

150. Tiedje, J. M., S. A. Boyd, and B. Z. Fathrepure. "Anaerobic Degradation of Chlorinated Aromatic Hydrocarbons," *Dev. Ind. Microbiol.* 27:117–127 (1987).

151. Fathrepure, B. Z., J. M. Tiedje, and S. A Boyd. "Reductive Dechlorination of Exachlorobenzene to Tri- and Dichlorobenzenes in Anaerobic Sewage Sludge," *Appl. Environ. Microbiol.* 54:327–330 (1988).

152. Bosma, T. N. P., J. R. van der Meer, G. Schraa, M. E. Tros, and A. J. B. Zehnder. "Reductive Dechlorination of all Trichloro- and Dichlorobenzene Isomers," *FEMS Microbiol. Ecol.* 53:223–229 (1988).

153. Mikesell, M. D., and S. A. Boyd. "Reductive Dechlorination of the Pesticides 2,4-D, 2,4,5-T and Pentachlorophenol in Anaerobic Sludges," *J. Environ. Qual.* 14:337–340 (1985).

154. Mikesell, M. D., and S. A. Boyd. "Complete Reductive Dechlorination and Mineralization of Pentachlorophenol by Anaerobic Microorganisms," *Appl. Environ. Microbiol.* 52:861–865 (1986).

155. Bryant, F. O., D. D. Hale, and J. E. Rogers. "Regiospecific Dechlorination of Pentachlorophenol by Dichlorophenol-Adapted Microorganisms in Freshwater, Anaerobic Sediment Slurries," *Appl. Environ. Microbiol.* 57:2293–2301 (1991).

156. Gibson, S. A., and J. M. Suflita. "Anaerobic Degradation of 2,4,5-Trichlorophenoxy-acetic Acid in Samples from Methanogenic Aquifer: Stimulation by Short-Chain Organic Acids and Alcohols," *Appl. Environ. Microbiol.* 56:1825–1832 (1990).

157. Brown, J. F., Jr., D. L. Bedard, M. J. Brennan, J. C. Carnahan, H. Feng, and R. E. Wagner. "Polychlorinated Biphenyl Dechlorination in Aquatic Sediments," *Science* 236:709–712 (1987).

158. Quensen, J. F., III, S. A. Boyd, and J. M. Tiedje. "Dechlorination of Four Commercial Polychlorinated Biphenyl Mixtures (Aroclors) by Anaerobic Microorganisms from Sediments," *Appl. Environ. Microbiol.* 56:2360–2369 (1990).

159. Nies, L., and T. M. Vogel. "Effect of Organic Substrates on Dechlorination of Aroclor 1242 in Anaerobic Sediments," *Appl. Environ. Microbiol.* 56:2612–2617 (1990).

160. Ye, D., J. F. Quensen, III, J. M. Tiedje, and S. A. Boyd. "Anaerobic Dechlorination of Polychlorobiphenyls (Aroclor 1242) by Pasteurized and Ethanol-Treated Microorganisms from Sediments," *Appl. Environ. Microbiol.* 58:1110–1114 (1992).

161. Adriaens, P., and D. Grbić-Galić. "Anaerobic Biodegradation of Highly Chlorinated Dioxins and Dibenzofurans," paper presented at the Dioxin '91 meeting, Research Triangle Park, Chapel Hill, NC, September 1991.

162. Reineke, W. "Microbial Degradation of Halogenated Aromatic Compounds," in *Microbial Degradation of Organic Compounds*, D. T. Gibson, Ed. (New York: Marcel Dekker Inc., 1984), pp. 319–360.

163. Karns, J. S., J. J. Kilbane, S. Duttagupta, and A. M. Chakrabarty. "Metabolism of Halophenols by 2,4,5-Trichlorophenoxyacetic Acid Degrading *Pseudomonas cepacia*," *Appl. Environ. Microbiol.* 46:1176–1181 (1983a).

164. Marks, T. S., R. Wait, A. R. W. Smith, and A. V. Quirk. "The Origin of the Oxygen Incorporated during the Dehalogenation/Hydroxylation of 4-Chlorobenzoic Acid by an *Arthrobacter* sp.," *Biochem. Biophys. Res. Commun.* 124:669–674 (1984).

165. Adriaens, P., H.-P. E. Kohler, D. Kohler-Staub, and D. D. Focht. "Bacterial Dehalogenation of Chlorobenzoates and Coculture Biodegradation of 4,4'-Dichlorobiphenyl," *Appl. Environ. Microbiol.* 55:887–892 (1989).

166. Crawford, R. L., and W. W. Mohn. "Microbiological Removal of Pentachlorophenol from Soil using a *Flavobacterium*," *Enzyme Microb. Technol.* 7:617–620 (1985).

167. Dorn, E. M., and H.-J. Knackmuss. "Chemical Structure and Biodegradability of Halogenated Aromatic Compounds: Two Catechol 1,2-Dioxygenases from a 3-Chlorobenzoate Grown Pseudomonad," *J. Biochem.* 174:73–84 (1978).

168. Hickey W. J., and D. D. Focht. "Degradation of Mono-, Di-, and Trihalogenated Benzoates by *Pseudomonas aeruginosa* JB2," *Appl. Environ. Microbiol.* 56:3842–3850 (1990).

169. Ahmed, M., and D. D. Focht. "Degradation of Polychlorinated Biphenyls by Two Species of *Achromobacter*," *Can. J. Microbiol.* 19:47–52 (1973).

170. Brunner, W., F. H. Sutherland, and D. D. Focht. "Enhanced Biodegradation of Polychlorinated Biphenyls in Soil by Analog Enrichment and Bacterial Inoculum," *J. Environ. Qual.* 14:324–328 (1985).

171. Kohler, H.-P. E., D. Kohler-Staub, and D. D. Focht. "Cometabolism of Polychlorinated Biphenyls: Enhanced Transformation of Aroclor 1254 by Growing Bacterial Cells," *Appl. Environ. Microbiol.* 54:1940–1945 (1988).

172. Adriaens, P., C.-M. Huang, and D. D. Focht. "Biodegradation of PCBs by Aerobic Microorganisms," in *Organic Substances in Sediments and Water, Vol 1*, R. Baker, Ed. (Chelsea, MI: Lewis Publishers Inc., 1991), pp. 311–326.

173. Adriaens, P., and D. D. Focht. "Cometabolism of 3,4-Dichlorobenzoate by *Acinetobacter* sp. Strain 4CB1," *Appl. Environ. Microbiol.* 57:173–179 (1991).

174. Janke, D., O. V. Maltseva, L. A. Golovleva, and W. Fritsche. "On the Relation Between Cometabolic Monochloroaniline Turnover and Intermediary Metabolism in *Rhodococcus* sp. An 117," *Z. Allg. Microbiol.* 24:305–316 (1984).

175. Bumpus, J. A., M. Tien, D. Wright, and S. D. Aust. "Oxidation of Persistent Environmental Pollutants by a White Rot Fungus," *Science* 228:1434–1436 (1985).

176. Aust, S. D. "Degradation of Environmental Pollutants by *Phanerochaete chrysosporium*," *Microb. Ecol.* 20:197–209 (1990).

177. Bumpus, J. A., and S. D. Aust. "Biodegradation of Environmental Pollutants by the White Rot Fungus *Phanerochaete chrysosporium*: Involvement of the Lignin-Degrading System," *Bioessays* 6:166–170 (1987).

178. Lamar, R. T., and D. M. Dietrich. "Use of White-Rot Fungi for the On-Site Remediation of Pentachlorophenol Contaminated Soils," paper presented at the International Conference on In-Situ and On-Site Bioremediation, San Diego, CA, March 1991.

179. Adriaens, P., and D. D. Focht. "Evidence for Inhibitory Substrate Interactions During Cometabolism of 3,4-Dichlorobenzoate by *Acinetobacter* sp. Strain 4-CB1," *FEMS Microbiol. Ecol.* 85:293–300 (1991).

180. Hernandez, B. S., F. K. Higson, R. Kondrat, and D. D. Focht. "Metabolism of and Inhibition by Chlorobenzoates in *Pseudomonas putida* P111," *Appl. Environ. Microbiol.* 57:3361–3366 (1991).

181. Nelson, W. O., and P. G. C. Campbell. "The Effects of Acidification on the Geochemistry of Al, Cd, Pb, and Hg in Freshwater Environments: A Literature Review," *Environ. Pollut.* 71:91–130 (1991).

182. Scheuhammer, A. M. "Acidification-Related Changes in the Biogeochemistry and Ecotoxicology of Mercury, Cadmium, Lead, and Aluminium," *Environ. Pollut.* 71:87–90 (1991).

183. Gilmour, C. C., and E. A. Henry. "Mercury Methylation in Aquatic Systems Affected by Acid Deposition." *Environ. Pollut.* 71:131–169 (1991).

184. Wood, J. M. "Environmental Pollution by Mercury," *Adv. Environ. Sci. Technol.* 2:36–56 (1971).

185. Silver, S., and T. G. Kinscherf. "Genetic and Biochemical Basis for Microbial Transformations and Detoxification of Mercury and Mercurial Compounds," in *Biodegradation and Detoxification of Environmental Pollutants*, A. M. Chakrabarty, Ed. (Boca Raton, FL: CRC Press, 1982), pp. 86–99.

186. Robinson, J. B., and O. H. Tuovinen. "Mechanisms of Microbial Resistance and Detoxification of Mercury and Organomercury Compounds: Physiological, Biochemical and Genetic Analyses," *Microbiol. Rev.* 48:95–124 (1984).

187. Wood, J. M. "Biological Cycles for Toxic Elements in the Environment," *Science* 183:1049–1052 (1974).

188. Edmonds, J. S., and K. A. Francesconi. "Transformation of Arsenic in the Marine Environment," *Experientia* 43:553–557 (1987).

189. Fleming, R. W., and M. Alexander. "Dimethylselenide and Dimethyltelluride Formation by a Strain of *Pennicilium*," *Appl. Microbiol.* 24:424–429 (1972).

190. Doran, J. W. "Microorganisms and the Biological Cycling of Selenium," in *Advances in Microbial Ecology, Vol. 6*, K. C. Marshall, Ed. (New York: Plenum Press, 1982), pp. 1–32.

191. Karlson, U., and W. T. Frankenberger Jr. "Accelerated Rates of Selenium Volatilization from California Soils," *Soil Sci. Soc. Am. J.* 53:749–753 (1989).

192. Huey, C., F. E. Brinkman, S. Grim, and W. P. Iverson. In *Proceedings of the International Conference on Transport of Persistent Chemicals in Aquatic Ecosystems*, Q. L. Laltham, Ed. (Ottawa, Canada: National Research Council of Canada, 1974), pp. II-73–II-78.

193. Wong, P. T. S., Y. K. Chau, and P. L. Luxon. "Methylation of Lead in the Environment," *Nature (London)* 253:263–264 (1975).

194. Ghiorse, W. C. "Biology of Iron- and Manganese-Depositing Bacteria," *Ann. Rev. Microbiol.* 38:515–550 (1984).

195. Ghiorse, W. C. "Microbial Reduction of Manganese and Iron," in *Biology of Anaerobic Microorganisms*, A. J. B. Zehnder, Ed. (New York: John Wiley and Sons, Inc., 1988), pp. 305–333.

196. Lovley, D. R. "Dissimilatory Fe (III) and Mn (IV) Reduction," *Microbiol. Rev.* 55:259–287 (1991).

197. Black, J. P. "Microbial Ecology Associated with Manganese-Binding Bacterial Exopolymers," Ph.D. Thesis, Harvard University, Cambridge, MA (1991).

198. Stevenson, F. J. *Cycles of Soil: Carbon, Nitrogen, Phosphorus, Sulfur, Micronutrients* (New York: Wiley-Interscience, 1986), p. 380.

199. Wuhrman, K. "Stream Purification," in *Water Pollution Microbiology*, R. Mitchell, Ed. (New York: Wiley-Interscience, 1972), pp. 119–151.

200. McLean, R. J. C., D. Beauchemin, L. Clapham, and T. J. Beveridge. "Metal-Binding Characteristics of the Gamma-Glutamyl Capsular Polymer of *Bacillus licheniformis* ATCC 9945," *Appl. Environ. Microbiol.* 56:3671–3677 (1990).

201. Chanmugathas, P., and J.-M. Bollag. "Microbial Role in Immobilization and Subsequent Mobilization of Cadmium in Soil Suspensions," *Soil Sci. Soc. Am. J.* 51:1184–1191 (1987).

202. Berman, M., T. Chase, Jr., and R. Bartha. "Carbon Flow in Mercury Biomethylation by *Desulfovibrio desulfuricans*," *Appl. Environ. Microbiol.* 56:298–300 (1990).

203. Oremland, R. S., C. W. Culbertson, and M. R. Winfrey. "Methylmercury Decomposition in Sediments and Bacterial Cultures: Involvement of Methanogens and Sulfate Reducers in Oxidative Demethylation," *Appl. Environ. Microbiol.* 57:130–137 (1991).

204. Doran, J. W., and M. Alexander. "Microbial Transformation of Volatile Se Compounds on Soil," *Soil Sci. Soc. Am. J.* 40:687–690 (1977).

205. Oremland, R. S., J. T. Hollibaugh, A. S. Maest, T. S. Presser, L. G. Miller, and C. W. Culbertson. "Selenate Reduction to Elemental Selenium by Anaerobic Bacteria in Sediments and Culture: Biogeochemical Significance of a Novel, Sulfate-Independent Respiration," *Appl. Environ. Microbiol.* 55:2333–2343 (1989).

206. Oremland, R. S., N. A. Steinberg, T. S. Presser, and L. G. Miller. "In Situ Bacterial Selenate Reduction in the Agricultural Drainage Systems of Western Nevada," *Appl. Environ. Microbiol.* 57:615–617 (1991).

207. Thompson-Eagle, E. T., W. T. Frankenberger, Jr., and U. Karlson. "Volatilization of Selenium by *Alternaria alternata*," *Appl. Environ. Microbiol.* 55:1406–1413 (1989).

208. Calderone, S. J., W. T. Frankenberger, Jr., D. R. Parker, and U. Karlson. "Influence of Temperature and Organic Amendments on the Mobilization of Selenium in Sediments," *Soil Biol. Biochem.* 22:615–620 (1990).

209. Anderson, L. C. D., and K. W. Bruland. "Biochemistry of Arsenic in Natural Waters: The Importance of Methylated Species," *Environ. Sci. Technol.* 25:420–427 (1991).

210. Cullen, W. R., and K. J. Reimer. "Arsenic Speciation in the Environment," *Chem. Rev.* 89:713–764 (1989).

211. Masscheleyn, P. H., R. D. Delaune, and W. H. Patrick, Jr. "Effect of Redox Potential and pH on Arsenic Speciation and Solubility in a Contaminated Soil," *Environ. Sci. Technol.* 25:1414–1419 (1991).

212. Ishibashi, Y., C. Cervantes, and S. Silver. "Chromium Reduction in *Pseudomonas putida*," *Appl. Environ. Microbiol.* 56:2268–2270 (1990).

213. Bopp, L. H., and H. L. Ehrlich. "Chromate Resistance and Reduction in *Pseudomonas fluorescens* Strain LB300," *Arch. Microbiol.* 150:426–431 (1988).

214. Gvozdyak, P. I., N. F. Mogilevich, A. F. Ryl'skii, and N. I. Grishchenko. "Reduction of Hexavalent Chromium by Collection Strains of Bacteria," *Mikrobiologiya* 55:962–965 (1986).

215. Wang, P.-C., T. Mori, K. Komori, M. Sasatsu, K. Toda, and H. Ohtake. "Isolation and Characterization of an *Enterobacter cloacae* Strain that Reduces Hexavalent Chromium under Anaerobic Conditions," *Appl. Environ. Microbiol.* 55:1665–1669 (1989).

216. Chanmugathas, P., and J.-M. Bollag. "A Column Study of the Biological Mobilization and Speciation of Cadmium in Soil," *Arch. Environ. Contam. Toxicol.* 17:229–237 (1988).

217. Silva-Avalos, J., M. G. Richmond, O. Nagappan, and D. A. Kunz. "Degradation of the Metal-Cyano Complex Tetracyanonickelate(II) by Cyanide-Utilizing Bacterial Isolates," *Appl. Environ. Microbiol.* 56:3664–3670 (1990).

218. Knowles, C. J., and A. W. Bunch. "Microbial Cyanide Metabolism," *Adv. Microb. Physiol.* 27:73–111 (1986).

219. Smith, M. R., J. L. Lequerica, and M. R. Hart. "Inhibition of Methanogenesis and Carbon Metabolism in *Methanosarcina* sp. by Cyanide," *J. Bacteriol.* 162:67–71 (1985).

220. Fedorak, P. M., D. J. Roberts, and S. E. Hrudey. "The Effects of Cyanide on the Methanogenic Degradation of Phenolic Compounds," *Water Res.* 20:1315–1320 (1986).

221. Fedorak, P. M., and S. E. Hrudey. "Cyanide Transformation in Anaerobic Phenol-Degrading Methanogenic Cultures," *Water Sci. Technol.* 21:67–76 (1989).

222. Fallon, R. D., D. A. Cooper, R. Speece, and M. Henson. "Anaerobic Biodegradation of Cyanide under Methanogenic Conditions," *Appl. Environ. Microbiol.* 57:1656–1662 (1991).

223. Strandberg, G. W., S. E. Shumate, II, J. R. Parrot, Jr., and S. E. North. "Microbial Accumulation of Uranium, Radium, and Cesium," NBS Special Publication (U.S.) 618:274–285 (1981).

224. Lovley, D. R., E. J. P. Phillips, Y. A. Gorby, and E. R. Landa. "Microbial Reduction of Uranium," *Nature (London)* 350:413–416 (1991).

225. Gorby, Y. A, and D. R. Lovley. "Enzymatic Uranium Precipitation," *Environ. Sci. Technol.* 26:205–207 (1992).

226. Lovley, D. R., and E. J. P. Phillips. "Reduction of Uranium by *Desulfovibrio desulfuricans*," *Appl. Environ. Microbiol.* 58:850–856 (1992).

227. Harvey, R. S., and R. Patrick. "Concentration of 137Cs, 65Zn, and 85Sr by Freshwater Algae," *Biotechnol. Bioeng.* 9:449–456 (1967).

228. Tomioka, N., H. Uchiyama, and O. Yagi. "Isolation and Characterization of Cesium-Accumulating Bacteria," *Appl. Environ. Microbiol.* 58:1019–1023 (1992).

229. Madsen, E. L., J. L. Sinclair, and W. C. Ghiorse. "In Situ Biodegradation: Microbiological Patterns in a Contaminated Aquifer," *Science* 252:830–833 (1991).

CHAPTER **6**

Genetic Strategies for Strain Improvement

B. D. Ensley

1. INTRODUCTION

Many toxic substances are decomposed or altered by microorganisms because that ability gives them an advantage in their natural environment. These organisms synthesize biologically active materials such as enzymes or polymers at rates, reactivities, and in an abundance that balances with other needs of the organisms in that particular ecological niche. Frequently, the desired activities are too slow or inefficient for widespread application by man in remediating highly contaminated sites. Beneficial traits of microorganisms such as the ability to synthesize amino acids or antibiotics have been improved by mutation and selection for decades. More recently, modern mutagens such as transposons and gene cloning techniques have come into use. These methods can also be applied to improving characteristics in microorganisms that destroy or detoxify hazardous materials.

There are a number of ways to introduce a desirable genetic change into a microorganism. A cell can be made to produce more of an enzyme so that wastes

0-87371-613-2/94/$0.00+$.50
© 1994 by Lewis Publishers

may be degraded more rapidly or to a greater extent. An enzyme may be changed to cause it to react with a broader range of substrates and degrade a variety of hazardous compounds. Sometimes a catalytic activity merely has to be nudged to be synthesized at all times to make a useful new organism. Changing the way a degradative pathway is regulated is an important advantage of using more sophisticated molecular biology techniques and will be discussed in the next section.

Strategies for genetic improvement of microorganisms can be as simple as exposing a bacterium to a mutagen like nitrogen mustard and screening on agar plates for survivors that appear to grow faster on an organic waste or display a higher resistance to a heavy metal. These genetically improved strains can then be subjected to additional rounds of treatment and screening. Such efforts are effective but also labor intensive and cumbersome. More predictable and potentially more useful changes can be made in the genetic code of a microorganism if the biochemistry and the molecular biology of the desired activity are characterized. In this case, rationally designed experiments can be planned and executed that will cause the target organism to display improved performance.

At the same time, genetically altered microorganisms that attack hazardous waste, unlike their cousins in the specialty chemical and pharmaceutical fermentation areas, will not be confined to aseptic reactors under optimum growth conditions. To be useful, these organisms will have to function in a hostile and competitive environment, degrading toxic waste in a variety of different conditions. The use of genetically altered cultures in hazardous waste treatment demands that we appreciate the effects of genetic changes on the ability of the organism to survive or compete. A microorganism debilitated by the methods used to alter it will not function as well in the environment and will soon be overcome. The altered organism must have the ability to function as a robust member in a changing microbial population.

2. ENHANCEMENT OF CATABOLIC ACTIVITY

Many organic hazardous wastes are degraded by microorganisms in a series of enzyme catalyzed reactions. Typically, the molecules are broken down to carbon dioxide and water. Approximately half of the organic carbon is incorporated into cellular constituents of microorganisms including new copies of the enzymes that broke down the waste in the first place. This assimilative process continues until some other limiting factor such as availability of trace nutrients or oxygen restricts growth. This type of metabolism works well with simple organic compounds that are not highly toxic and are easily converted by enzymes adapted for their biocatalysis. Improved catabolic activity may be necessary to efficiently degrade the complex and highly toxic organic compounds present in chemical synthetic process streams and chlorinated wastes. Polychlorinated biphenyls, chloroaromatics, and high molecular weight polycyclic aromatic hydrocarbons are examples of hazardous organic molecules that could be destroyed more rapidly and to a greater extent by microorganisms with improved catabolic potential. Microorganisms with the ability to rapidly and completely degrade these types of wastes will broaden applications in the biological treatment of contaminated sites.

2.1 Enhanced Uptake of Substrates

With only a few exceptions, such as oxidation by extracellular lignin-degrading enzymes from *Phanerochaete chrysosporium*,[1] hazardous waste molecules can only be degraded if they are transferred to the interior of the cell. Degradative enzymes are usually not secreted by the organisms; many require the presence of cofactors for activity, and most of them are relatively unstable in the extracellular environment. Consequently, uptake or transport of hazardous waste is an essential precondition for degradation. Some hazardous wastes, particularly charged molecules, can be transferred into the cell through active transport processes. Transport proteins are synthesized to concentrate substrate molecules inside the cell and are either directly reactive with the target hazardous waste itself or with close analogs. For example, fatty acids can be concentrated up to tenfold inside the cell against a concentration gradient.[2] In *Escherichia coli*, the fad genes encode proteins involved in transport of a range of fatty acids from C_{11} to C_{18} and show some activity against C_7 to C_{10} molecules as well indicating that this uptake system has a broad substrate specificity. The medium chain length fatty acids can serve as substrates for the fad enzymes, but cannot induce their synthesis. Mutants constitutive in the synthesis of the uptake proteins could be one example of an improvement in substrate transport.[2] As with other enzyme systems involved in hazardous waste degradation, altering an enzymatic uptake system to broaden the specificity or to increase the rate of uptake by mutagenesis or cloning techniques will improve performance if this element is rate limiting. Overproduction of an uptake system with a mutant or hybrid promoter also presents possible increased uptake potential, although transport proteins are by nature membrane-bound or membrane-associated, and experience has taught that over-expression of membrane-bound protein can be difficult.

Many of the most recalcitrant and notorious hazardous wastes have little or no surface charge and do not appear to be substrates for active transport mechanisms. The route of uptake for these molecules, which contain only carbon, hydrogen, and sometimes a few halogen atoms, appears to be almost entirely due to equilibrium partitioning between the bulk water phase and the relatively hydrophobic microorganisms. Most industrial solvents such as chloroform, dichloromethane, and trichloroethylene as well as polychlorinated biphenyls (PCBs) and polycylic aromatic hydrocarbons fit into this category. Their very recalcitrance may be due in part to low water solubility that contributes to poor biological availability. Using genetic methods to enhance uptake of these substrates is both more complicated than altering active transport processes and more likely to result in significantly enhanced rates of degradation since bioavailability appears to be a major limiting factor in the degradation of many insoluble hazardous wastes.

One approach to the improved uptake of hydrophobic organics is to increase the synthesis of molecules that solubilize or emulsify target substrates. The growth rate of bacteria on relatively insoluble substrates has been shown to be directly related to the solubility of the molecules in water[3] suggesting that emulsification aids degradation since only substrates in the dissolved state are taken up by microorganisms. This has also been shown by studies with naphthalene and biphenyl as substrates,[4] where microbial growth rates were found to be dependent solely on the soluble concentrations and unaffected by larger amounts of insoluble hydrocarbon.

Certain microorganisms such as *Acinetobacter calcoaceticus* RAG-1 synthesize potent extracellular emulsifying agents that are polysaccharide or lipopolysaccharide in nature.[5,6] These emulsifiers act on a range of different molecules including the complex mixtures of hydrocarbons found in kerosene and gasoline.[7] Complex extracellular polymers containing proteins, lipids, and carbohydrates with emulsifying activity against hydrocarbons are elaborated by species of *Corynebacterium* and *Acinetobacter* during growth on alkanes.[8,9] Other complex trehaloselipids[10] and rhamnolipids[11] are also synthesized by microorganisms during growth on alkanes. The role of rhamnolipids in stimulating growth and paraffin degradation has been demonstrated through the use of mutants capable of only limited growth with paraffin as a sole carbon source. Addition of purified rhamnolipids to these cultures markedly stimulated growth at the expense of n-paraffins illustrating the natural role of these bioemulsifiers.

It is highly likely that steps taken to increase the bioavailability of insoluble hazardous wastes would significantly contribute to their removal from the environment. However, modifying several genes to overproduce the complex mixtures of proteins, carbohydrates, and lipids elaborated by microorganisms as bioemulsifiers will be difficult. It may be simpler to introduce synthetic emulsifiers or surfactants that provide the same performance at much less effort and ultimately less cost.

Genetic approaches to improving the uptake of actively transported molecules include (1) broadening the substrate specificity of the uptake mechanism, (2) causing uptake protein synthesis in the absence of a natural inducer, or (3) increasing the number of uptake molecules or proteins in the cell. Improved rates or extent of degradation could follow in those instances where uptake is the rate limiting step. Many of the most widespread and persistent hazardous wastes do not appear to be transported into cells by enzyme catalyzed mechanisms. Genetic alterations that improve or introduce the synthesis of bioemulsifiers or direct addition of synthetic detergents may impact uptake rates for these molecules.

2.2 Modification of Catabolic Enzyme Systems

Degradation or detoxification of complex hazardous wastes often depends upon a series of reactions catalyzed by several enzymes in a complete system. Modification of these systems can mean (1) improving the efficiency and extent of degradation, (2) changing the flow of metabolites through the degradative pathway, or (3) broadening the substrate specificity of the enzymes. Sometimes, more than one improvement can be introduced into the same pathway by using a combination of the techniques available.

2.2.1 Broadened Substrate Specificity

Of all the genetic improvements that can be made to a microorganism used in the treatment of hazardous waste, one of the most promising is the synthesis of enzymes or complete enzyme systems that display the broadest range of substrates. This approach can make a single organism active against a wide variety of wastes and helps address the issue that biodegradation of a few or even most contaminants may not be sufficient for a site to be regarded as completely cleaned. The original patent issued for a genetically modified organism[12] describes degradation of a variety of hydrocarbon wastes by a single microorganism.

There are two related methods for broadening substrate specificity using genetic techniques. The first is by the direct introduction of one or more mutations into a gene encoding the desired enzyme. A new enzyme is synthesized that is able to attack a wider range of organic molecules. An organism having an enzyme system active against benzene, for example, may be modified by mutagenesis to also attack toluene, xylene, or related compounds. Site-directed mutagenesis, the use of recombinant DNA techniques to precisely change one or more amino acids in a protein sequence, has recently been employed to broaden the substrate specificity of the enzyme subtilisin.[13,14] Specific amino acid changes were introduced into the subtilisin molecule that were calculated to increase the number of substrates that could be attacked by this enzyme. This rational approach to protein engineering through site-directed mutagenesis was successful in significantly broadening the activity of subtilisin and demonstrating the potential practical application of this technique in many areas of enzyme catalysis.

If the DNA sequence and biochemistry of a pathway are less understood, the technique of randomly introducing mutations into a gene or simply selecting for organisms that display activity against a broader range of substrates can be used. Expanding the biodegradative activity of *Pseudomonas citronellolis* against recalcitrant branched hydrocarbons through the use of enzyme recruitment has been reported.[15] Introduction of a metabolic plasmid by conjugation into *P. citronellolis* produced an extended growth substrate range. Better results were obtained with the same organism by using random mutagenesis and strong selection for constitutive expression of an alkane degradative pathway. The modified organism was now able to degrade 2, 6-dimethyl-2-octane as a growth substrate. This complex hydrocarbon could be oxidized by the existing biochemical pathways in the organism, but was not an inducer. Constitutive enzyme synthesis permitted growth of the organism with a broader substrate range.

Conjugation has also been used to construct a microorganism that has degradative capabilities against 3-chlorobenzoate, 4-chlorobenzoate, and 3, 5-dichlorobenzoate. The original organism, able to grow only with 3-chlorobenzoate as its sole carbon and energy source, was modified by introducing the TOL plasmid from *Pseudomo-*

nas putida mt-2 into the original *Pseudomonas* sp. B13.[16] It was later shown that the new organism used enzymes both from the introduced TOL plasmid and chromosomally encoded enzymes for the degradation of molecules such as 4-chloro- and 3, 5-dichlorobenzoates. While both plasmid and chromosomally encoded enzymes were responsible for early steps in the degradation of 4-chloro- and 3, 5-dichlorobenzoate, only the catechol oxygenase from the parent strain was active since the plasmid-encoded meta-cleavage pathway is inactivated by the chlorocatechol products.[17]

Subjecting indigenous microorganisms to strong selective pressure in a chemostat by forcing growth at the expense of highly recalcitrant compounds such as 2, 4, 5-trichlorophenoxyacetic acid (2, 4, 5-T) can also be used to enrich for new activities. An organism, arising presumably from natural recombination and random mutagenesis events, was isolated from this type of chemostat culture within a 10-month period.[18,19] The utility of this approach has been limited, however, by the fact that most organic compounds are not completely degraded by a single enzyme. If each enzyme along the pathway is not mutated to attack new substrates in concert, the modified organism will not be able to use the new compound for growth and cannot be isolated by selection. For example, a benzene dioxygenase may be modified by mutagenesis to also attack toluene to form a diol or cresol product. The second enzyme in the pathway must also be able to attack the new intermediate substrate and so on down the chain of reactions. Success depends upon a series of complimentary mutations, and the chances of achieving this through classic mutagenesis and selection techniques are not good.

Transferring genes from one organism to another to increase substrate range through the use of transposons or gene cloning methods offers a rational approach for pathway engineering and has been anticipated by workers dealing with the problem of chloroaromatic degradation.[20] The metabolic pathways involved in hydrocarbon metabolism have been divided by Timmis and co-workers[21] into a series of "throats" and "stomachs". These authors have surmised that combinations of "throats" and "stomachs" can be transferred into various organisms by the use of transposons or plasmids to construct new organisms with a broader potential for degradation. It was proposed that a common set of intermediary biochemical pathways with broad specificity could be used as the "stomachs" for a variety of "throats" capable of attacking most or all of the common organic hazardous wastes.

A combination of techniques including random mutagenesis and gene cloning with plasmids and transposons has been used to construct microorganisms with broad substrate specificity against a variety of chloroaromatics.[20] This work also illustrates that the same approaches can be used to achieve several objectives at once: the new organism has a pathway engineered to avoid synthesis of toxic (to the organism) intermediates; the metabolic pathway displays activity against a much broader range of substrates than it did originally, and a series of recruited enzymes form the basis of a complete biochemical pathway for a family of chloroaromatics found in process waters.

The above constructs are prototypes for genetic enhancement methods used to improve the properties of a hazardous waste degrading microorganism. Other examples of these techniques are appearing in literature with greater frequency. Recently a paper describing the cloning, nucleotide sequence, and expression in a variety of gram negative organisms of the gene encoding the enzyme haloalkane dehalogenase was published.[22] This enzyme has a broad substrate specificity with activity against chloromethanes and ethanes, dichloroethanes and propanes, and bromo and iodoalkanes.[23] This enzyme can be efficiently expressed from its own promoter sequences in a variety of gram negative bacteria including halocarboxylic acid degraders such as *Pseudomonas* sp. GJ1, *Xanthobacter autotrophicus* XD, and *Pseudomonas* sp. GJ31.[24,25] Introduction of this plasmid-encoded haloalkane dehalogenase permitted *Pseudomonas* GJ1 to degrade 1, 2-dichloroethane with complete dechlorination and allowed *Xanthobacter autotrophicus* XD and *Pseudomonas* GJ31 to utilize 1-chlorobutane as a sole carbon and energy source for growth. The use of this single gene as a new "throat" permitted the already existing biochemical pathways of these organisms to metabolize the reaction products and thus considerably broaden the range of substrates that could be attacked. Other such examples of enzyme recruitment will appear in the literature as this method of genetically altering catabolic enzyme systems becomes better understood and applied.

2.2.2 *Increased Stability*

All catabolic enzymes have a defined activity lifetime. This lifetime may be considerably reduced during degradation of marginal or nongrowth substrates including chlorinated molecules. Destruction of catalytic activity can occur and should be avoided if at all possible, because without the necessary catalytic enzymes, the desired degradative reactions will rapidly grind to a halt. Since the organism and the researchers have gone to so much trouble to synthesize these catalysts, it is reasonable to assume that any modification that can increase stability will improve the performance of the entire process. If the compound causing the enzyme inactivation cannot be avoided by pathway engineering, then the enzyme itself must be modified to increase its stability, or a more stable enzyme must be found.

There are examples of enzyme inactivation during the oxidation of a gratuitous substrate that is also a hazardous waste. The rapid inactivation of the methane monooxygenase enzyme system during oxidation of trichloroethylene (TCE) has been documented.[26] Trichloroethylene also inactivates the toluene dioxygenase system from *Pseudomonas* sp. F1.[27] Because TCE is such a widespread environmental contaminant, development of enzyme systems with more resistance to inactivation by this substrate would be highly useful. Increasing the stability of an enzyme by random mutagenesis and selection or screening is theoretically possible, but designing and implementing an efficient strategy is extremely difficult since pleomorphic effects of mutations can often mask the desired results (e.g., a mutant enzyme that still attacked toluene or methane with reduced activity against TCE would appear more stable).

More successful in some instances has been the rational redesign of enzymes or enzyme systems based on information regarding the amino acid sequences involved in enzyme stability. Recently, the technique of site directed mutagenesis has been used to markedly increase the stability of the commercially useful enzyme subtilisin.[28,29] By changing amino acid residues subject to inactivation by temperature or proteolysis, the stability the subtilisin molecule was increased by an order of magnitude. While this particular enzyme has limited application in hazardous waste degradation, it serves as an example of what can be done using modern genetic strategies for improving the stability of enzymes that have cloned and well characterized genes.

Increased stability of a multi-component enzyme system involved in the degradation of a polycyclic aromatic hydrocarbon has also recently been achieved. The naphthalene dioxygenase enzyme system gratuitously oxidizes many substrates including the molecule indole. Naphthalene dioxygenase-catalyzed oxidation of indole causes inactivation of the enzyme system with this normally robust and sturdy catalytic system displaying a half-life of from $1^1/_2$ to 2 h during the oxidation of indole. A combination of techniques including supplementing the reaction medium, genetically amplifying the unstable component of the multi-component system, and using site-directed mutagenesis designed to stabilize the catalytic iron in one component resulted in an order of magnitude increase in enzyme half-life.[30] If similar results can be obtained with enzyme systems that catalyze the oxidation of polychlorinated biphenyls, polycyclic aromatic hydrocarbons, or chlorinated solvents, the improved stability will markedly increase the efficiency of degradation. This approach is currently limited to well studied pathways. The naphthalene dioxygenase enzyme system could be stabilized because there was a thorough understanding of the biochemistry and molecular biology of naphthalene oxidation.

2.3 Modified Synthesis of Catabolic Enzymes

Biosynthesis of desired catabolic enzymes can be genetically modified to improve the performance of the host organism. Among the most common modifications is genetic alterations that result in the constitutive synthesis of enzymes or complete enzyme systems through simple mutagenesis and selection. Constitutive synthesis of catabolic enzymes is a desirable modification because some hazardous wastes that otherwise would be degraded by microorganisms do not induce the synthesis of the necessary enzymes and therefore persist in the environment. If an enzyme system can be synthesized in the absence of its natural inducer, the enzymes may gratuitously degrade other molecules. Recently the genes encoding the toluene monooxygenase enzyme system from *Pseudomonas cepacia* strain G4, which will oxidize TCE, were altered by the use of transposon mutagenesis to cause constitutive synthesis.[31] This mutant was active against TCE and dichloroethylene in the absence of an aromatic inducer emphasizing the simplicity and flexibility such altered organisms offer. The constitutive synthesis of toluene monooxygenase is but one example of this useful genetic alteration of microorganisms that degrade hazardous wastes.

The enzyme system responsible for the degradation of polychlorinated biphenyls (PCBs) has been cloned onto plasmids and transferred into *Escherichia coli* where modified synthesis in the form of constitutive expression is observed.[32] The PCB degrading genes from *Pseudomonas* strain LB400, a microorganism known to catabolize a very wide variety of PCB congeners, were isolated and transferred onto plasmids that replicate in *E. coli*. Measurement of PCB degrading activity in the recombinant *E. coli* demonstrated that the enzymes responsible for this activity are constitutively synthesized, either from a promoter carried on the plasmid or from a promoter contained within the cloned DNA fragment. As a consequence of this constitutive PCB degrading activity, the recombinant *E. coli* displayed activity against PCB even when grown under noninducing conditions. The original *Pseudomonas* strain must grow in the presence of biphenyl or another inducer for maximum PCB degrading activity and has much lower enzyme activity after growth in the absence of an inducer. The recombinant *E. coli* has improved enzyme synthesis in the sense that an inducer is no longer required for maximum PCB degradative activity. Enzyme activity of strain FM4560 may be further improved through the use of regulated, high-level expression directed by a recombinant promoter and operator region.

There are several examples of the use of hybrid promoter and operator sequences for regulated expression of recombinant catabolic enzymes. Different application needs have been met using genetic engineering techniques. The toluene monooxygenase enzyme system from *Pseudomonas mendocina* has been cloned into several plasmids that provide regulated expression in *Escherichia coli*. This altered regulation provides distinct advantages over wild-type expression. The enzyme system in the wild-type microorganism is unstable in the absence of a natural inducer such as toluene (Figure 1). Enzyme activity is rapidly lost from the culture if toluene is removed. Activity of the enzyme system in the recombinant organism is significantly improved. Altered regulation of toluene monooxygenase is beneficial in the degradation of nongrowth substrates such as trichloroethylene (TCE) by this enzyme system.[33] Because synthesis of this enzyme system requires the presence of toluene in the natural isolate, degradation of TCE and other chlorinated ethenes is subject to competitive inhibition by toluene. Both the chlorinated ethene and the natural substrate compete for the same active site on toluene monooxygenase. The recombinant microorganism does not require toluene, so there is no competitive inhibition effects against TCE, and as a consequence, more efficient and extensive degradation is observed. In this example, the recombinant microorganisms do not degrade either toluene or TCE any faster than the wild-type organism does, but the altered regulation permits synthesis of the enzyme in the absence of interfering substrates and allows more efficient and extensive degradation to proceed.

Sometimes the gene encoding an enzyme is cloned or otherwise genetically altered simply to cause overproduction of the desired enzyme. Cloning and

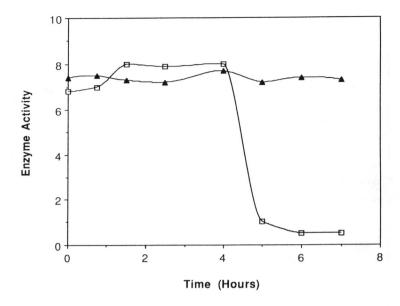

Figure 1. Lifetime of toluene monooxygenase activity in the absence of inducer. *Pseudomonas mendocina* strain KR1 was grown in the presence of toluene, and a recombinant *Escherichia coli* harboring the TMO genes was grown and the activity induced with a temperature induction cycle as described previously.[33] Cells from both cultures were then harvested separately and incubated in basal salts medium plus glutamate. The cells were sampled at frequent intervals, and the toluene monooxygenase activity measured. While the TMO activity in the recombinant *E. coli* (▲) is stable for periods in excess of 24 h, TMO activity in the wild-type organism (□) declines by at least 90% after 4 h of incubation in the absence of the toluene inducer.

genetic manipulation can take the form of adding a new, stronger promoter or actually altering the DNA sequence of the gene itself. Recently, an enzyme involved in 4-chlorobenzoate dehalogenation has been cloned and over-expressed on a plasmid in *E. coli*. While over-expression of proteins is often used to enhance enzyme activity, in this case the expression of the cloned genes was used to elucidate the biochemical pathway by which this enzyme acts against chlorobenzoates.[34] The cloning itself aided in the study and characterization of the enzyme: something which had proven to be very difficult using the wild-type microorganism. This approach resulted in successful dissection of the dehalogenation pathway and identified unusual dehalogenation cofactors. Molecular biology techniques were effective tools in elucidating the biochemistry of this dehalogenation process and may aid in improving the performance of this enzyme system.

Modified synthesis of catabolic enzymes can also result in substantial overproduction of a desired enzyme. The enzyme parathion hydrolase is involved in the hydrolysis of a number of organo-phosphate pesticides.[35] This enzyme is synthesized by a number of microorganisms including *Pseudomonas diminuta*, a *Flavobacterium* species, and a recombinant *Streptomyces*. Attempts to express the cloned genes in *E. coli* originally resulted in only low level synthesis of the enzyme. N-terminal amino acid sequence analysis of the peptide from the wild-type microorganism revealed, in combination with DNA sequence analysis of the cloned gene, that 29 amino acids were removed during synthesis of the mature periplasmic protein. Since *E. coli* lacked the necessary protein processing machinery, a new parathion hydrolase gene was constructed in which the first 29 amino acid residues were deleted to directly encode the mature protein and a new ribosome binding site and start codon were inserted in front of the structural gene. This new, genetically matured gene was then retransformed into *E. coli*, and the activity was examined after induction. While *Pseudomonas diminuta* would synthesize approximately two units of enzyme per milligram of whole cell protein, the recombinant *E. coli*, after induction of the strong P_L promoter, contained 50 units of enzyme per milligram of crude cell protein.[36] Overproduction of this enzyme was not limited to the specific activity of units per cell. Since the growth behavior of *E. coli* is so well characterized, this organism can be grown to extremely high densities, and soluble crude extract activities over 100-fold higher than those from microorganisms such as *Streptomyces* could be obtained.[37] The modified synthesis of this enzyme illustrates several consequences of genetic manipulation. A strong promoter and operator region was used to significantly increase synthesis of the protein, and the processing machinery present in the original host cell was substituted by genetic processing. The power of molecular biology techniques in improving gene expression is apparent in this example.

3. ENHANCEMENT OF INORGANIC TRANSFORMATIONS

Microbial treatment of inorganic contaminants usually involves the immobilization or removal of metals and metal salts. Microorganisms can alter the chemical characteristics of metals and particularly their abundance in the environment by mechanisms such as precipitation, accumulation inside the cells, alteration in the valence state of the metal through oxidation or reduction reactions, and transformation by methylation and demethylation. Many microbial metal transformation processes are systems evolved by microorganisms to provide resistance against toxicity. Microbial processes can, in some cases, treat two types of metal contaminants. Some heavy metals such as mercury or lead are directly toxic, while other metals are radioactive. Biologically synthesized extracellular polymers can be used to treat either type.

All of the various processes used by microorganisms to interact with metals may be amenable to improved performance through genetic manipulation. Enzymes involved in precipitation, uptake, valence changes, or methylation of heavy metals can be manipulated genetically by amplification. In addition, microbial synthesis and elaboration of complex molecules such as polysaccharides can remove metals by chelation to these extracellular polymers. The synthesis and binding properties of these polymers can also be improved through genetic manipulation.

3.1 Synthesis of Extracellular Polymers

Polymers such as capsular slime on cell walls can bind heavy metals. The capsular slime of both gram positive and gram negative bacteria binds a variety of toxic heavy metals including chromium, cobalt, nickel, and copper.[38] While heavy metals can be concentrated out of solution onto cell walls, membranes, and polysaccharide capsules of bacteria, genetic changes to improve the performance of metal binding by bacteria cannot, given today's technology, alter appreciably the characteristics of a bacteria's cell wall or membrane surface. However, genetic changes in the synthesis of extracellular polysaccharides may be useful in improving heavy metal binding by bacteria.

Slime and capsular extracellular polymers of microorganisms are active anions under natural conditions. The anodic nature of many polysaccharides is caused by carboxylic acid sidechains such as uronic, pyruvic, and syalic acids. The prototypical polysaccharide synthesizing microorganism is *Zoogloea ramigera* strain 115. Under certain growth conditions, this organism can convert 60% of the available growth substrate into water soluble exopolysaccharide containing approximately 3 to 5% pyruvate. The negatively charged carbonic groups of the pyruvate are thought to be primarily responsible for the polymer's metal ion affinity.[39] The polysaccharide from *Z. ramigera* can remove Co, Cu, Fe, and Ni from solution, while mutants that do not produce the polymer bind significantly less of these metals.[40] Other capsule producing microorganisms also bind metals. Noncapsulated strains of *Enterobacter aerogenes* bind significantly less Cu, Cd, Mn, and Co than similarly encapsulated strains, and six- to tenfold more Cu and Cd is bound by capsular polymers than is absorbed by the cells.[41]

Extracellular polymers and crude microbial biomass have also been used to bind metals that are radioactive. Cells of *Penicillium chrysogenum* and sludge from a municipal wastewater treatment plant are effective absorbers of radium.[42] Bacterial polymers have also been used to remove plutonium from water. Immobilized cells of *Pseudomonas aeruginosa* can concentrate PuO_2 or $PuCl_4$ from a minimal medium solution.[43] Some extracellular polymer processes for removing radionuclides have been taken to the pilot stage. Whole cells of the fungus *Rhizopus arrhizus* have been used in a pilot scale reactor demonstrating the absorption and concentration of uranium, thorium, aluminum, and yttrium bound to the microbial biomass.[44]

The gross synthesis of extracellular polysaccharides can be altered by growth conditions in some strains. Synthesis of polysaccharides by *Xanthomonas* is directly proportional to growth rate with polymers produced by rapidly growing cells having both a different structure and total bulk. The synthesis of polymers by *Pseudomonas atlantica* is enhanced by increased surface area for growth and stimulated under conditions that cause cells to be stressed such as growth in the stationary phase.[45,46] Because the synthesis of extracellular polymers can be dependent upon external conditions and the presence of specific pathways for carbohydrate synthesis, manipulation of growth conditions or the use of mutagenesis or genetic engineering techniques could improve the biological absorption of metals from aqueous environments. Studies describing the cloning, artificial regulation, or over-expression of genes involved in exopolysaccharide biosynthesis have not yet been published. When these methods are developed, synthesis of metal binding polymers by microorganisms could markedly improve economics of bioabsorption and expand the range of applications that are competitive with chemically synthesized polymers and other metal removal methods.

3.2 Structure of Extracellular Polymers

Polymers elaborated by microorganisms can take many forms. Factors controlling the structure of extracellular polysaccharides are not well understood, and these molecules are not readily altered using current genetic techniques. In other instances, such as secreted enzymes or peptides, alteration in polymer structure by genetic techniques is straightforward and well characterized. There are a number of metal binding proteins including ferritin, metallothionenes, heme proteins, and other proteinaceous metal chelators. Recent genetic engineering of a metal binding site in a protein has been reported[47] providing opportunities for speculation regarding mutated or genetically engineered proteins being used to absorb heavy metals. In all but the most specialized cases, where extremely low final metal concentrations are required, proteins as heavy metal adsorbents will most likely not be commercially competitive with biologically or chemically synthesized polysaccharides or other adsorbents. Because of the relatively high molecular weight of even small proteins, a protein with many metal binding sites per molecule will still absorb less metal on a weight basis than a simple polysaccharide.

Genetic changes that affect the structure of extracellular polysaccharides, while more difficult to perform, promise to yield substantial benefit in increasing the efficiency of metal binding by microorganisms. The number of metal ions that can be bound by a unit of polysaccharide mass is influenced by the number of negatively charged groups on the polysaccharide. Just as has been observed with the bulk of exopolysaccharides synthesized, growth conditions can affect the structure of polysaccharides elaborated by microorganisms such as *Xanthomonas campestris*.[45] Growth conditions also affect the composition of the exopolymer produced by *Pseudomonas atlantica*,[46] and the content of uronic, galacturonic, and glucuronic acids in the extracellular polysaccharide of this organism is altered by growth rate.[48]

Changes in the composition of exopolymers caused by manipulating culture conditions suggest that engineering of biosynthetic pathways can also be used to affect the density of acidic groups on extracellular polysaccharides. Once the genes involved in controlling the structure or composition of acidic side-chains have been identified, alterations in the regulation or expression level of these genes could result in the synthesis of new polymeric materials with higher binding capacities or lower binding constants. While random mutagenesis and selection techniques may be successful in improving heavy metal binding by extracellular polymers, direct genetic intervention through recombinant DNA methods will be used for improved synthesis and structure of polymers. A step in this direction has already been taken by workers at the Massachusetts Institute of Technology. Mutants of the microorganism *Zoogloea ramigera* have been isolated that synthesize a new exopolysaccharide and have the ability to take up foreign DNA with conventional transformation techniques.[39] These mutant strains of *Z. ramigera* provide two useful tools for genetic enhancement of metal binding exopolysaccharide structure. First, the polysaccharide secreted by these mutants displays a 30 to 50% increase in the pyruvate content of the molecule. Since this mutation has a direct impact on the synthesis of a pyruvate rich polysaccharide, which should have increased metal binding properties, identification of this gene is simplified, and the mutation itself is useful. Secondly, this organism provides a genetic tool because, unlike the parent, it can take up DNA after simple and well characterized treatments. The use of this new strain as a host for polysaccharide biosynthetic genes will permit study of gene expression in both *Z. ramigera* and other exopolysaccharide synthesizing microorganisms and has potential for producing new polymers by introducing novel genetic combinations into this organism.

3.3 Enzymes for Biogenic Mineral Transformation

Certain biological reactions cause alterations in the chemical state of many heavy metals and could be manipulated to display increased activity in metal transformation. The biochemical activities of certain anaerobic microorganisms are an example. Production of hydrogen sulfide through the activity of sulfate-reducing bacteria causes the formation of hydrogen sulfide (H_2S), which reacts with some metals to form insoluble sulfides. Metal precipitation by sulfate-reducing bacteria occurs in natural environments.[49] These natural environments have been emulated with impoundments for treating process waters from mining operations.[50] Manipulation of culture conditions or genes involved in sulfate reduction could stimulate this process of precipitation.

Another method of removing metals from solution by precipitation is the formation of phosphates derived from a phosphatase enzyme located on the surface of certain bacteria. Microorganisms such as *Citrobacter* can use phosphatase to cause a precipitation of cadmium, lead, and uranium. An immobilized biofilm of this organism has been used to remove uranium from water.[51,52] While the conversion of sulfate to H_2S by sulfate-reducing microorganisms is a primary metabolic activity and as such may be difficult to enhance through alterations in genetic information, overproduction of a single enzyme such as phosphatase can be easily accomplished. This enzyme depends upon glycerol 2-phosphate as a phosphate source, so synthesis of this intermediate by the microorganisms may also have to be enhanced to give an altered enzyme-substrate system providing higher extracellular phosphatase activity. In any case, there is ample evidence suggesting that synthesis of simple molecules such as glycerol phosphate will not be a rate-limiting step, and overproduction of this single enzyme may considerably increase the rate and perhaps the extent of metal removal by extracellular phosphatase.

Some metals are actively transported across the bacterial membrane. The specific system may be used for the uptake of an essential element, and gratuitous transport processes can result in substantial accumulation of heavy metals that are not otherwise metabolically significant.[53] Transport processes cause high level silver accumulation by microorganisms such as *Pseudomonas*.[54] Uranium, radium, and cesium[55] are also transported and accumulated in microorganisms. Genetic alterations caused by random mutagenesis, the use of transposons, or genetic engineering techniques that enhance heavy metal uptake offer a promising way to remove heavy metals from the immediate environment. A problem remains, however, of how to recover biomass that now contains high accumulations of heavy metals. It is not advantageous for a microorganism to take up heavy metals if this removal is temporary and the metals are released when the organism dies or decomposes. To be useful, heavy metal uptake processes must also be coupled with methods of recovering the microorganism or chemically altering the heavy metal and its abundance in the environment.

Changes in the valence state of various metals are mediated by biologically catalyzed oxidation and reduction reactions. The formation of insoluble oxides of manganese has been demonstrated with some bacterial species.[56] Changes in the valence of other heavy metals have also been reported including the reduction of molybdenum and gold by microorganisms. Perhaps the best understood of all microbial-heavy metal interactions is a combined active transport and reductive process involved in microbial mercury resistance. These well-studied systems offer the greatest current potential for enhancement through genetic manipulation techniques.

One mechanism of microbial mercury resistance is encoded by a small operon, called the mer operon, carrying genes for mercuric reductase and mercury uptake whose expression is induced by mercury. This enzyme system catalyzes the reduction of Hg (II) to mercury metal Hg (0), which is volatile, and causes removal of the mercury from the local environment. The first step in mercury metabolism is an active transport, mercury-inducible mechanism comprising three to four genes.[57] An active transport system for mercury can cause accumulation of this toxic heavy metal in the cells and in instances where either the accompanying mercuric reductase activity has been deleted by cloning or mutation, or the over-expression of uptake proteins causes excessive mercury uptake, the cells display a mercury hypersensitive phenotype.[58] Once mercury is transported into the cell, it becomes a substrate for mercuric reductase (the mer A gene product). The inducible NADPH-dependent mercuric reductase converts Hg (II) to Hg (0) causing the volatilization of this metal.

Concerted amplification of both the mercury uptake system and the mercuric reductase has significant potential for improved performance. Genetic enhancement to broaden activity against other heavy metals and high-level synthesis of the enzymes in the absence of mercury offer considerable promise. Mutations and direct genetic engineering techniques can accomplish changes designed to improve the enzymatic metabolism of heavy metals.

4. REFERENCES

1. Lamar, R. T., M. J. Larson, and T. K. Kirk. "Sensitivity to and Degradation of Pentachlorophenol by *Phanerochaete* spp.," *Appl. Environ. Microbiol.* 56:3519–3526 (1990).

2. Maloy, S. R., C. L. Ginsburgh, R. W. Simons, and W. D. Nunn. "Transport of Long and Medium Chain Fatty Acids by *Escherichia coli* K12," *J. Biol. Chem.* 256:3735–3742 (1981).

3. Woodzinski, R. S., and M. Johnson. "Yields of Bacterial Cells from Hydrocarbons," *Appl. Microbiol.* 16:1886–1891 (1968).

4. Woodzinski, R., and D. Bertollini. "Physical State in Which Naphthalene and Bibenzyl Are Utilized by Bacteria," *Appl. Microbiol.* 23:1077–1081 (1972).

5. Reisfeld, A., E. Rosenburg, and D. Gutnick. "Microbial Degradation of Crude Oil: Factors Effecting the Dispersion of Sea Water by Mixed and Pure Cultures," *Appl. Environ. Microbiol.* 24:363–368 (1972).

6. Zuckerberg, A., A. Diver, Z. Peeri, D. L. Gutnick, and E. Rosenburg. "Emulsifier of Arthrobacter RAG-1: Chemical and Physical Properties," *Appl. Environ. Microbiol.* 37:414–420 (1979).

7. Rosenburg, E., A. Perry, D. T. Gibson, and D. L. Gutnick. "Emulsifier of Arthrobacter RAG-1: Specificity of Hydrocarbon Substrate," *Appl. Environ. Microbiol.* 37:409–413 (1979).

8. Kappeli, O., and W. R. Finnerty. "Partitioning of Alkane by an Extracellular Vesicle-Derived from Hexadecane Grown *Acinetobacter*," *J. Bacteriol.* 140:707–712 (1979).

9. Zajic, J. E., H. Guigmard, and D. F. Gerson. "Emulsifying and Surface Active Agents from *Corynebacterium hydrocarboclastus*," *Biotechnol. Bioeng.* 29:1285–1301 (1977).

10. Suzuki, T., K. Tanaka, I. Matsubara, and S. Kinoshita. "Trehalose Lipids and Alpha-Branched Beta Hydroxene Fatty Acid Formed by Bacterial Growth on n-Alkanes," *Agr. Biol. Chem.* 33:1619–1627 (1969).

11. Itoh, S., and T. Suzuki. "Effect of Rhamnolipids on Growth of a *Pseudomonas aeruginosa* Mutant Deficient in n-Paraffin-Utilizing Ability," *Agr. Biol. Chem.* 36:2233–2235 (1972).

12. Chakrabarty, A. M. "Microorganisms Having Multiple Compatible Degradative Energy Generating Plasmids and Preparation Thereof," U.S. Patent No. 4,259,444 (1981).

13. Wells, J. A., B. C. Cunningham, T. P. Graycar, and D. A. Estell. "Recruitment of Substrate Specificity Proteins from One Enzyme into a Related One by Protein Engineering," *Proc. Natl. Acad. Sci. U.S.A.* 84:5157–5174 (1987).

14. Wells, J. A., D. B. Bowers, R. R. Bott, T. P. Graycar, and D. A. Estell. "Designing Substrate Specificity by Protein Engineering of Electrostatic Interactions," *Proc. Natl. Acad. Sci. U.S.A.* 84:1219–1223 (1987).

15. Fall, R., J. Brown, and T. L. Schaeffer. "Enzyme Recruitment Allows the Biodegradation of Recalcitrant Branched Hydrocarbons by *Pseudomonas citronellolis*," *Appl. Environ. Microbiol.* 38:715–722 (1979).

16. Reineke, W., and H.-J. Knackmuss. "Construction of Haloaromatics Utilizing Bacteria," *Nature* 277:385–386 (1979).

17. Reineke, W., and H.-J. Knackmuss. "Hybrid Pathway for Chlorobenzoate Metabolism in *Pseudomonas* sp. B13 Derivatives," *J. Bacteriol.* 142:467–473 (1980).

18. Ghosal, D., D. Chatterjee, and A. M. Chakrabarty. "Microbial Degradation of Halogenated Compounds," *Science* 228:135–142 (1985).

19. Chakrabarty, A. M., J. S. Karns, J. Kilbane, and D. K. Chatterjee. "Selective Evolution of Genes for Enhanced Degradation of Persistent, Toxic Chemicals," in *Genetic Manipulations: Impact on Man in Society*, W. Arbor, K. Illmensee, W. J. Peacock, and P. Starlinger, Eds. (Miami, FL: ICSU Press, 1984), pp. 43–54.

20. Timmis, K. N., F. Rojo, and J. L. Ramos. "Prospects for Laboratory Engineering of Bacteria to Degrade Pollutants," in *Environmental Biotechnology: Reducing Risks from Environmental Chemicals Through Biotechnology*, G. S. Omen, Ed. (New York: Plenum Press, Inc., 1988), pp. 61–79.

21. Timmis, K. N., P. R. Lehrbach, S. Harayama, R. Dom, N. Marmaude, S. Bass, R. Leppik, A. Weightman, W. Reineke, and H.-J. Knackmuss. "Analysis and Manipulation of Plasmid Encoded Pathways for the Catabolism of Aromatic Compounds by Soil Bacteria," in *Plasmids in Bacteria*, D. R. Helinski, S. N. Cohen, D. B. Clewell, D. A. Jackson, and A. Hollaender, Eds. (New York: Plenum Press, Inc., 1985), pp. 719–739.

22. Janssen D. B., F. Pries, J. Van der Ploeg, B. Kazemier, P. Terpstra, and B. Witholt. "Cloning of 1, 2-Dichloroethane Degradation Genes of *Xanthobacter autotrophicus* GJ10 and Expression and Sequencing of the dhl A Gene," *J. Bacteriol.* 171:6791–6799 (1989).

23. Keuning, S., D. B. Jenssen, and B. Witholt. "Purification and Characterization of Hydrolytic Haloalkane Dehalogenase from *Xanthobacter autotrophicus* GJ10," *J. Bacteriol.* 163:635–639 (1985).

24. Janssen, D. B., A. Scheper, and B. Witholt. "Biodegradation of 2-Chloromethane and 1, 2-Dichloroethane by Pure Bacterial Cultures," *Progr. Ind. Microbiol.* 20:169–178 (1984).

25. Oldenhuis, R., L. Kuijk, A. Lammers, D. B. Janssen, and B. Witholt. "Degradation of Chlorinated and Non-Chlorinated Aromatic Solvents in Soil Suspension by Pure Bacterial Cultures," *Appl. Microbiol. Biotechnol.* 30:211–217 (1989).

26. Fox, B. G., J. G. Borneman, L. Wackett, and J. Lipscomb. "Haloalkane Oxidation by the Soluble Methane Monooxygenase from *Methylosinus trichosporium* OB-3b: Mechanistic and Environmental Implications," *Biochemistry* 29:6419–6427 (1990).

27. Wackett, L., and S. Householder. "Toxicity of Trichloroethylene to *Pseudomonas putida* F1 Is Mediated by Toluene Dioxygenase," *Appl. Environ. Microbiol.* 55:2723–2725 (1989).

28. Narhi, L. L., Y. Stabinski, M. Levitt, L. Miller, R. Saghdev, S. Finley, S. Park, C. Kolvenbach, T. Arakawa, and M. Zukowski. "Enhanced Stability of Subtilisin by 3 Point Mutations," *Biotechnol. Appl. Biochem.* 13:12–24 (1991).

29. Zukowski, M., Y. Stabinski, L. Narhi, J. Mauck, M. Stowers, and N. Fiske. "An Engineered Subtilisin with Improved Stability: Applications in Human Diagnostics," in *Genetics and Biotechnology of the Bacilli, Volume 3*, M. Zukowski, A. T. Ganesan, and J. A. Hoch, Eds. (New York: Academic Press, 1990).

30. Serdar, C., D. Murdock, and B. Ensley. "Enhancement of Naphthalene Dioxygenase Activity During Microbial Indigo Production," World Patent Office No. 9102055 (1991).

31. Shields, M. S. "Construction of a *Pseudomonas cepacia* Strain Constituitive for the Degradation of Trichloroethylene and its Evaluation for Field and Bioreactor Studies," *Abstr. Annu. Meet. Am. Soc. Microbiol.* p. 215 (1991).

32. Mondello, F. J. "Cloning and Expression in *Escherichia coli* of *Pseudomonas* Strain LB400 Genes Encoding Polychlorinated Biphenyl Degradation," *J. Bacteriol.* 171:1725–1732 (1989).

33. Winter, R. B., K. M. Yen, and B. D. Ensley. "Efficient Degradation of Trichloroethylene by a Recombinant *Escherichia coli*," *Bio/Technology* 7:282–286 (1989).

34. Scholten, J. D., K.-H.Chang, C. Babbitt, H. Charest, M. Sylvestre, and D. Dunaway-Marriano. "Novel Enzymic Hydrolytic Dehalogenation of a Chlorinated Aromatic," *Science* 253:182–185 (1991).

35. Munnecke, D. M., and D. P. Hsieh. "Pathways of Microbial Metabolism of Parathion," *Appl. Environ. Microbiol.* 31:63–69 (1976).

36. Serdar, C. M., D. C. Murdock, and M. F. Rohde. "Parathion Hydrolase Gene from *Pseudomonas diminuta* MG: Subcloning, Complete Nucleotide Sequence and Expression of the Mature Portion of the Enzyme in *Escherichia coli*," *Bio/Technology* 7:1151–1155 (1989).

37. Steiert, G. J., B. M. Pogell, M. K. Speedie, and J. Lorado. "A Gene Coding for a Membrane-Bound Hydrolase Is Expressed as a Secreted, Soluble Enzyme in *Streptomyces lividans*," *Bio/Technology* 7:65–68 (1989).

38. McLean, R. J., and T. J. Beveridge. "Metal-Binding Capacity of Bacterial Surfaces and Their Ability to Form Mineralized Aggregates," in *Microbial Mineral Recovery*, H. L. Ehrlich and C. L. Brierley, Eds. (New York: McGraw Hill Publishing Company, 1990), pp. 185–221.

39. Easson, D. D., O. P. Peoples, and A. J. Sinskey. "*Zoogloea* Transformation Using Exopolysaccharide Non-Capsule Producing Strains," U.S. Patent No. 4,948,733 (1990).

40. Friedman, B. A., and P. R. Dugan. "Concentration and Accumulation of Metallic Ions by the Bacterium *Zoogloea*," *Dev. Ind. Microbiol.* 9:381–394 (1968).

41. Rudd, T., R. M. Sterritt, and J. M. Bluster. "Mass Balance of Heavy Metal Uptake by Encapsulated Cultures of *Klebsiella aerogenes*," *Microb. Ecol.* 9:261–270 (1983).

42. Tsezos, M., and D. M. Keller. "Adsorption of Radium-226 by Biological Origin Adsorbents," *Biotechnol. Bioeng.* 25:201–215 (1983).

43. Meyer, H. R., J. E. Johnson, R. P. Tengerby, and P. M. Goldman. "Use of a Bacterial Polymer Composite to Concentrate Plutonium from Aqueous Media," *Health Phys.* 37:359–363 (1979).

44. Tsezos, M., and R. G. L. McCready. "The Pilot Plant Testing of the Continuous Extraction of the Radionuclides Using Immobilized Biomass," in *Environmental Biotechnology for Waste Treatment*, G. S. Sayler, Ed. (New York: Plenum Press, 1991), pp. 249–260.

45. Tait, M. I., I. W. Sutherland, and A. J. Clarke-Sturman. "Effect of Growth Conditions on the Production, Composition, and Viscosity of *Xanthomonas campestris* Exopolysaccharide," *J. Gen. Microbiol.* 132:1483–1486 (1986).

46. Uhlinger, D. J., and D. C. White. "Relationship Between Physiological Status and Formation of Extracellular Polysaccharide Glycocalyx in *Pseudomonas atlantica*," *Appl. Environ. Microbiol.* 45:64–68 (1983).

47. Arnold, F. H., and B. L. Haymore. "Engineered Metal-Binding Proteins: Purification to Protein Folding," *Science* 252:1796–1797 (1991).

48. Gorden G., E. Quintero, and G. Geesey. "Effect of Dilution Rates on *Pseudomonas atlantica* Exopolymer," *Bacteriol. Abst.* (1988).

49. Jackson, T. A. "The Biogeochemistry of Heavy Metals in Polluted Lakes and Streams at Flin Flon, Canada," *Environ. Geol.* 2:173–189 (1978).

50. Brierley, J., and C. L. Brierley. "Biological Methods to Remove Selected Inorganic Pollutants from Uranium Mine Waste Water," in *Biogeochemistry of Ancient and Modern Environments*, P. A. Trudringer, M. R. Walter, and B. J. Ralph, Eds. (Canberra, Australia: Australian Academy of Science, 1980), pp. 661–667.

51. Macaskie, L. E., and A. C. Dean. "Uranium Accumulation by a *Citrobacter* sp. Immobilized as Biofilm on Various Support Materials," in *Proceedings Fourth European Congress on Biotechnology*, O. N. Neijssel, R. R. Van der Mere, and K. C. H. Luyben, Eds. (Amsterdam: Elsevier, 1987), pp. 37–40.

52. Macaskie, L. E., and A. C. Dean. "Uranium Accumulation by Immobilized Biofilms of a *Citrobacter* sp.," in *Biohydrometallurgy, Science and Technology Letters*, P. R. Norris and D. P. Kelly, Eds. (U.K.: New Surrey, 1988), pp. 556–557.

53. Brierley, C., D. P. Kelly, K. J. Seal, and D. J. Best. "Materials and Biotechnology," in *Biotechnology Principles and Applications*, I. J. Higgins, D. J. Best, and J. Jones, Eds. (Oxford: Blackwell Scientific, 1985), pp. 163–212.

54. Charlie, R. C., and A. T. Bull. "Bioaccumulation of Silver by a Multi-Species Community of Bacteria," *Arch. Microbiol.* 123:239–244 (1979).

55. Stranburg, G. W., S. E. Shoemade, J. R. Parrott, and S. E. North. "Microbial Accumulation of Uranium, Radium and Cesium," in *Environmental Speciation and Monitoring Needs for Trace Metal-Containing Substances from Energy-Related Processes*, F. E. Brinkman and R. H. Fish, Eds. (Washington, DC: U.S. Department of Commerce National Bureau of Standards, 1981), p. 27.

56. Mann S., N. H. Sparks, G. H. Scott, and E. W. de Vrind-de Jong. "Oxidation of Manganese and Formation of Mn_3, O_4 by Spore Coats of a Marine *Bacillus* sp.," *Appl. Environ. Microbiol.* 54:2140–2143 (1988).

57. Summers, A. O. "Organization, Expression, and Evolution of Genes from Mercury Resistance," *Ann. Rev. Microbiol.* 40:607–634 (1986).

58. Makahara, H., S. Silver, T. Miki, and R. H. Rownd. "Hypersensitivity to Hg^{2+} and Hyperbinding Activity Associated with Cloned Fragments of the Mercurial Resistance Operon of Plasmid MR1," *J. Bacteriol.* 140:161–166 (1979).

CHAPTER 7

Engineering Considerations

Graham Andrews

1. INTRODUCTION

1.1 Objectives and Methods of Bioprocessing Engineering

The first question to be answered is why this book has a chapter on engineering considerations. What are the distinctive contributions of the biochemical/sanitary/environmental engineering professions to the development of bioremediation technologies? What do engineers do, and how do they do it?

Engineering is sometimes defined as the design, construction, and operation of things that work, but this is inexact. Many competent auto mechanics could assemble the parts to build a functioning car. Similarly, many microbiologists and fermentation technologists could put together a working bioremediation process based mainly on other bioprocesses they have seen. This must not be confused with engineering. The concern of the engineer is not so much the many possible process configurations that will work, but the single one that works best. Engi-

0-87371-613-2/94/$0.00+$.50
© 1994 by Lewis Publishers

neering is essentially an economic optimization problem, "best" being defined as fulfilling the treatment objectives at the lowest possible total cost.

The quantities that must be optimized are the process variables, defined as any quantity that is under the direct control of the process designer and operator. One of the difficulties with *in situ* bioremediation is that there are few process variables: only the type and location of the wells and the flow, composition, and concentration of the nutrient/bacteria stream injected into them. Attempting to accommodate the many geochemical, hydrological, and microbiological complexities of an *in situ* process with so few adjustable parameters is a major challenge. The design of surface bioreactors is easier. Process variables include the type and size of the bioreactor, which fix the capital cost, the airflow, and addition of supplementary nutrients and power inputs for mixing and heating, which control the operating costs.

To establish the optimum values of these process variables, the engineer has three tools: experience, experiment, and mathematical modeling. Their relative importance varies from case to case. For example, the activated sludge process for treatment of municipal sewage is several decades old, and sewage varies little between cities. Previous experience is, therefore, a major element in the design of new plants. For more innovative bioprocesses such as the cometabolic degradation of chlorinated solvents in a fluidized-bed bioreactor, there is no such accumulated experience, and greater reliance must be put on experiment and modeling. *In situ* bioremediation is an extreme case. The indigenous microflora, geochemistry, and hydrology that determine the success or failure of the process are so site-specific that experience gained at one site may be useless at another.

The process called scale-up is actually a judicious integration of the three tools of engineering design. The philosophical basis of scale-up comes from Reynolds Similarity Principle, which says that two physical situations at different scales are identical if all of the relevant dimensionless numbers are the same. The nature of the bioprocess design problem is such that the formal procedures of Dimensional Analysis are rarely needed. Nevertheless, results should be expressed in dimensionless, or at least scale-independent, variables. Contaminant inflow for example is described by the loading (mass inflow per hour per unit bioreactor volume), which will be similar at the laboratory and commercial scale.

The rationale for the scale-up process is economic. Building and operating a commercial scale bioprocess are very expensive, so it is essential that the equipment works at optimum efficiency immediately after start-up. This is accomplished by optimizing the values of the process variables on smaller scale (pilot or laboratory) equipment, where experiments can be done much cheaper, and the consequences of failure are less disastrous. These experiments serve to validate mathematical models of the process, and the models are then used to design the equipment for experiments at the next larger scale. The number of intermediate scales over which this must be repeated varies between processes. It must be sufficient to get from the laboratory to commercial operation, while increasing the scale (i.e., bioreactor volume) by a factor of approximately 100 at each step.

1.2 Objectives of this Chapter

The starting point for doing scale-up is the quantitative understanding of bioprocesses derived from the simple mathematical models found in text books. These analyses have two essential functions. First, they allow an initial selection of bioreactor type. There are so many types available that it is impossible to test them all for a given application. Bioreactor analysis, coupled with knowledge of the specific microbe/contaminant system, allows a short list to be made from which the optimum type can be selected. For example, an inhibitory substrate dictates the choice of some type of completely mixed bioreactor.

Second, bioreactor analysis sets up "ideal bioreactors" against which the performance of real bioreactors can be judged. Consideration of these theoretical bioreactors allows us to identify those process variables critical to process performance and to avoid waste of experimental effort studying unimportant variables. The mean cell residence time in the activated sludge process is an excellent example of a process variable whose importance could not be guessed without mathematical analysis. The fact that the productivity (degradation rate per unit volume) of a properly designed aerobic bioreactor is usually controlled, not by the inherent microbial kinetics, but by the oxygen transfer rate is another example of the type of insight available from simple models. The fact that anaerobic processes suffer no such limitation explains why modern anaerobic bioreactors like the Upflow Anaerobic Sludge Blanket Reactor[1] have such high productivities despite their slower inherent microbial kinetics. The need for supplementation of the waste with additional nutrients is another area in which analysis can provide a starting point for experiments.[2]

Existing textbook analyses[3] are based on the extensive application of bioremediation processes in the treatment of municipal and industrial wastewaters. There has been considerable analysis of processes like activated sludge and the trickling filter that have been in use for many decades. Whereas, optimization of newer processes such as fluidized-bed bioreactors,[4] anaerobic filters,[1] and rotating biological contactors[5] has received less attention. Furthermore, process analysis has not kept up with the expanded opportunities for bioremediation offered by modern biotechnology and the challenges to bioprocess design offered by modern hazardous waste disposal problems. The objective of this chapter is to remedy this situation. Section 2 reviews the modern challenges to bioremediation process design; Section 3 examines the fundamentals of bioprocess modeling, and Section 4 shows how the insights obtained from this analysis can be applied to the challenges. The remainder of Section 1 will describe some of the differences between the scientific and engineering viewpoints with regard to process development.

1.3 Science and Engineering

Close collaboration between microbiologists and engineers is essential for the development of innovative bioremediation processes, and the earlier this collaboration begins the better. Too many promising developments have been hindered by microbiologists who thought they could do engineering and engineers who thought they could do microbiology. It is important that this collaboration is not inhibited by the differences in objectives and methods of the two professions. The following list of differences is not comprehensive, but may help to alleviate mutual incomprehension.

1.3.1 Nutrient Requirements

To the microbiologists, nutrient requirements are a matter of biochemistry. If yeast extract, for example, is needed by a particular strain, it will be added. To the engineer, nutrient requirements are a matter of economics. The cost of adding yeast extract on a commercial scale can destroy the economic feasibility of a process.

1.3.2 Mathematical Modeling

Integrating experiments with mathematical modeling is essential to successful scale-up of a process. Microbiologists are not usually trained in modeling and tend to resist the idea that something as complex as microbial metabolism can be reduced to a simple set of equations. While this is scientifically correct, it is irrelevant to the engineer who must have a tool to predict the performance of the process at the next larger scale. An approximate mathematical model is infinitely better than guesswork, particularly since the approximations can be covered by a "factor of safety" built into the design of the commercial equipment. It follows that a process development experiment whose goal is to obtain mathematical parameters for scale-up may be very different from an experiment whose objective is purely scientific. It is, for example, very difficult to extract information on the inherent microbial kinetics from the results of a batch culture experiment.

1.3.3 Timing of the Work

Science stresses building the experimental equipment quickly because the important work, the gathering and analysis of experimental data, can only start after it is built. Engineering is the exact opposite. The essential act is correct design of the equipment such that success is achieved the first time, as specified by the process requirements and predicted by the mathematical modeling. All of the design work happens before the equipment is built. This apparent difference between the scientific and engineering approach is most pronounced on a commercial scale. While scientists talk of the desirability of field-scale experiments, the idea is anathema to the engineer. If adjustments to the process variables are needed at field scale in order to meet the process requirements, then the engineering design process has failed. Such adjustments are difficult (pumps, for example, are not variable-speed at field scale), and the cost is likely to be met by the engineer being sued for incompetence. This is another reason for the safety factors built into the design of commercial-scale equipment.

1.3.4 Microbial Culture Composition

In nonsterile, mixed-culture bioprocesses, the composition of the microbial culture is not a process variable as defined above. Species composition of the culture is determined by competition and other ecological interactions among the microorganisms present at start-up of the process and those that subsequently enter the process with the waste and from the environment. The process designer and operator can affect the outcome only indirectly by adjusting process variables such as pH, temperature, and nutrient addition. As long as the waste provides an essential growth nutrient, species composition is of little concern; the species that consume the waste fastest and use it most efficiently for growth will dominate the culture. This inherent natural selection is what has allowed generations of sanitary engineers to design biological sewage treatment plants with surprisingly little input from microbiology. There is, however, an unfortunate tendency for this design approach to be carried over to cometabolic degradation processes, where natural selection produces no such guarantee of the survival of the most desirable species.

2. CHALLENGES FOR BIOREMEDIATION PROCESS DESIGN

Three modern trends are converging to present a major challenge to the designers of bioremediation processes. Public concern over hazardous chemicals in the environment is increasing the number and type of contaminants that must be treated. At the same time, advances of modern biotechnology are increasing the number and type of contaminants that can, potentially, be treated biologically. Finally, increasingly stringent regulations are setting higher standards for process performance and reliability. This section examines some differences between modern bioremediation problems and conventional applications to sewage and industrial wastewater treatment and discusses some of the necessary changes in bioreactor design.

2.1 Concentration Range

Conventional biological wastewater treatment processes work in a relatively narrow concentration range of organic matter from a few hundred mg/L for municipal sewage to a few g/L for concentrated industrial wastes. High organic loading has a profound effect on process design because the biological oxygen demand exceeds the solubility of oxygen from air in water (8 mg/L). The need to continuously transfer sufficient oxygen dictates the design of the bioreactors. In contrast, the objective in treating chlorinated solvents such as trichloroethylene (TCE) in groundwater is to reduce the concentration from a few mg/L down to drinking water standards of a few μg/L. Achieving these very low concentrations in a bioprocess requires an intimate knowledge of the microbiology and biochemistry involved and a new generation of bioreactors in which oxygen transfer is less critical because the oxygen dissolved within the water is sufficient for the entire process. These reactors, however, may need to combine biodegradation with physical removal processes such as adsorption on activated carbon in order to consistently produce an effluent of drinking water quality.

At the other extreme, bioprocessing is being considered for the degradation of pure nonaqueous organic waste, where incineration is not possible for regulatory or technological reasons. The effective organic concentrations of these wastes are in the order of 1000 g/L, far greater than their solubility in water. Bioprocessing in a nonaqueous phase, considered a long-range possibility for these wastes, would eliminate difficulties encountered with more conventional bioprocesses. The main difficulty is the formation of emulsions of the nonpolar organics as a nonaqueous phase within an aqueous phase than contains the microorganisms.[6] Emulsification produces many complications, and the kinetics of mass-transfer and biodegradation under these conditions are not well understood.

2.2 Volatility

Many contaminants that can potentially be treated biologically are both volatile and slightly soluble in water. The mixture of hydrocarbons in gasoline is a good example. Gasoline-contaminated air streams such as those generated by soil-venting operations can be treated in biofilters.[7] Gasoline-contaminated water can also be bioremediated, but the aeration required to stimulate aerobic microbial activity inevitably strips the hydrocarbons from the liquid, thus exchanging a water pollution problem for an air pollution problem.[8] The biodegradation of TCE by methanotrophic bacteria provides a more complex example of the same problem.[9] The contaminant is volatile; the process is aerobic and requires another gas, methane, as the carbon/energy source for the bacteria. The process design problem in all of the cases involves the interaction between gas/liquid mass-transfer and microbial kinetics. When considering this problem, it is tempting to rely on the considerable expertise that has accumulated on oxygen transfer into bioreactors. This approach is often misleading. The objective in aeration is to transfer oxygen at the highest possible rate. The fraction of the oxygen in the air entering the reactor that is actually transferred to the liquid and used by the microbes is irrelevant and is actually 2 to 3% in conventional bioreactors. A biofilter that removed only 2 to 3% of the contaminant fed into it would be an environmental disaster. A TCE bioremediation process that used only 2 to 3% of the methane fed into it would be an economic disaster, since methane, unlike air, is expensive.

Many process options exist to deal with gas mass-transfer and volatilization problems. Stripping of contaminants from the liquid can be reduced by minimizing airflow using active control of dissolved oxygen. Bubble columns in which the gas approximates plug flow while the liquid is completely mixed is one example of an appropriate bioreactor. Co-current flow of gas and liquid through a series of reactors or through a trickle bed reactor should also be considered. Other alternatives include the use of pure oxygen instead of air to eliminate the problem altogether by eliminating the off-gas and the pulsing of methane into a process instead of a continuous feed.

2.3 Solids

The emulsions mentioned in Section 2.1 are one example of the need for three-phase (air, water, nonaqueous liquid) bioreactors. Another, with solids replacing the nonaqueous liquid, is the bioremediation of soils and sludges. There are two main bioreactor configurations depending on the solids particle size. Fine particles such as clay soils are treated in mechanically-agitated or airlift-type slurry reactors. Coarse, dense particles such as sandy soils, which would be difficult to slurry, are treated in heaps or piles, a process known variously as composting or landfarming. Oxygen transfer is a critical process variable in either case.[10] The presence of solid particles reduces the oxygen transfer rate into a bubble-aerated slurry reactor. Oxygen enters heaps of solids by diffusion through the interstitial space, so the depth to which it can penetrate depends strongly on the fraction of this space filled by gas rather than liquid. Clay soils cannot be aerobically composted because the heap becomes waterlogged preventing oxygen penetration.

2.4 Inorganic and Mixed Wastes

Those responsible for sludge disposal from municipal treatment plants have long known (to their dismay) that biomass is an effective adsorbent for heavy metals such as cadmium and lead. Several companies are now selling this "biosorption" as a treatment technology for removing metals from wastewaters.[11] Microbial metabolism can also change the oxidation/reduction state of metals in the environment. Suitable choices of organisms and substrates can alter metals to a soluble form to facilitate soil washing or to an insoluble form for removal by precipitation.

Another role of inorganic compounds in modern bioremediation technology is as anaerobic electron acceptors. High nitrate wastewaters from the nuclear industry have been treated using denitrifying bacteria.[12] Mine drainage wastewaters high in metal sulfates are being treated in "engineered wetlands" in which sulfate-reducing bacteria play a critical role.[13] Denitrifying and sulfate-reducing bacteria have also been shown to degrade chlorinated solvents.[14] One of the advantages of bioremediation is the potential for carrying out several of these processes simultaneously. One can, for example, envision a groundwater contaminated with lead, nitrate, and chloroform. A bioprocess based on denitrifying bacteria may be able to reduce the nitrate to nitrogen, degrade the chloroform, and biosorb the lead simultaneously.

The use of these alternative electron acceptors frees the process designer from the rate limitations imposed by the need to transfer oxygen. Different bioreactor configurations can be considered, and very high concentrations of active biomass can be achieved by cell immobilization or recycling. This allows high reactor productivity despite the slow inherent microbial kinetics. Start-up times are, however, correspondingly increased.

Another role of inorganic compounds that must be anticipated by the bioprocess designer is that of inhibitory products. The biodegradation of explosives, for example, liberates large amounts of nitrogen as ammonia or nitrate[15] and that of TCE and similar compounds produces chloride. Any of these may, in an improperly designed process, reach inhibitory concentrations.

2.5 Cometabolic Processes and Recombinant Organisms

A final hurdle to the application of modern biotechnology to waste treatment is the fact that pure culture microbiology plays little role in conventional biological wastewater treatment. There are two reasons for this. First, it is economically impractical to seal commercial scale bioreactors and sterilize the waste in order to exclude undesired strains. Second, even if this were possible, it would be undesirable when the waste provides a major growth nutrient for the microorganisms. New microbial strains are constantly appearing in a nonsterile bioreactor, either introduced with the waste stream or from the environment or produced by spontaneous mutation or natural gene transfer. Those strains that can play a role in contaminant degradation with higher rates and growth yields than existing strains naturally dominate the culture. Others are washed out. Given time and a reasonably consistent waste stream, this natural selection usually produces the mixed culture best adapted to the particular mixture of contaminants. The exceptions usually involve process-specific physical requirements for the culture. For example, a desirable strain may wash out of the activated sludge process because it cannot attach to flocs in order to settle and be recycled. Or it may cause the flocs to form "bulking sludge", which upsets the process by not settling.

The rule that selection pressures favor process performance does not extend to cometabolic processes. A new microbial strain appearing in the bioreactor may grow faster on the available nutrients and thus take over the culture even though it is completely unable to degrade the target contaminant. Again consider the degradation of TCE by methanotrophic bacteria as an example. The methane monooxygenase enzyme that actually degrades TCE exists in two forms,[16] a soluble form active on TCE and an inactive particulate form. A contaminant strain containing the particulate form may, under certain conditions, have a competitive advantage because it does not waste its enzyme activity catalyzing reactions that do not benefit the cell. It may, therefore, take over a culture unless it could be rigorously excluded. Solutions are available for this problem in nonsterile bioreactors, but they call for levels of process control far exceeding present practice.

A related issue involves the use of genetically engineered microorganisms (GEMs) in waste treatment processes. One argument in response to the concern raised by the public and regulatory agencies about the release of such organisms into the environment is the inability of these microorganisms to compete with wild-type strains in the natural environment. The GEMs are weakened strains that will do their job in the short-term and then disappear. However, a weakened strain would not survive long in a nonsterile continuous bioreactor. The regulatory requirements concerning nonweakened strains are unclear, but the sterilization of the bioreactor effluent to prevent their release would be as difficult as sterilizing the influent. An effective pesticide-degrading organism may be welcome at a hazardous waste site, but not in an agricultural field. Considerable ingenuity and the close collaboration of geneticists and bioprocess engineers will be needed to resolve this difficulty.

A related problem for GEMs arises from the genetic instability of the recombinant organism and the possibility that the culture may become dominated by faster growing back-mutants that have lost the desired genetic trait. This type of problem has been resolved in the pharmaceutical industry, but only for batch processes where the time scale is so short that a back-mutant does not have time to dominate. Biological waste treatment is usually a continuous process where time is unlimited.

3. BIOPROCESS MODELING

3.1 Objectives and Tools of Modeling

Each of the challenges listed above requires some innovation in bioprocess design. What is the best (i.e.,cheapest) type of bioreactor for the desired level of treatment? What are the main process variables and their optimum values for steady-state operation? How should the process be controlled; that is, how should the process variables be adjusted in response to changes in the flow, strength, or composition of the waste? These questions cannot be answered based on experience because for a truly innovative process, previous experience is necessarily a poor guide. Nor can they be answered solely by experiment because there are so many variables that the experiments would be too time consuming and expensive. Questions must be answered by a properly integrated program of mathematical modeling and experiments on different scales. The models provide the quantitative insight into how the process is expected to work, identify the main process variables, and narrow the scope of experiments. The experiments check and refine each component of the model and provide parameter values needed for design of the full-scale process.

Bioprocess models are not meant to be exact mathematical descriptions of reality, an impossible complex objective in any field. Models are based, instead, on clearly stated assumptions about microbial metabolism, mass-transfer, mixing, etc. The objective is to include the main features of these phenomena and their interactions in order to determine process performance.

In setting up a model, it is essential to keep track of the types of equations involved, what aspects of the problem they describe, and how confidently they can be used. Rigorous conclusions must only be drawn from equations known to be a good description of reality. A complete model always consists of three sets of equations that differ markedly in the aspects of the process they describe and the rigor of the assumptions that underlie them.

3.1.1 Mass Balance Equations

The mass balance equations describe the principle of the conservation of mass for the microorganisms, each nutrient (including the waste), and microbial product. The form of these equations is determined by the type of bioreactor, and the only assumptions that must be made in their derivation concern the mixing conditions in the bioreactor. Mass balance equations can be written rigorously for plug-flow, complete mix, and various types of dispersion models. The only uncertainty concerns the ability of these models to describe actual large-scale bioreactors.

3.1.2 Yield Equations

Yield equations describe the relationship between the consumption of the contaminant, the consumption of other nutrients including electron acceptors, and the formation of biomass and other products. These equations are based on intracellular mass and energy balances and are quite rigorous if used correctly. Yield equations are critical for deciding which nutrient is limiting, a vital question in considering how process rates can be increased. The main uncertainty in their formulation lies in deciding which nutrients and processes are significant. A good example of the use of yield equations is the IAWPRC model for the activated sludge process,[17] which includes the separate processes of biomass growth, colloid adsorption and hydrolysis on the floc particles, biomass lysis, and nitrification. This model has a correspondingly large number of yield equations expressed in the form of a matrix.

3.1.3 Rate Equations

Rate equations attempt to give a quantitative description of the rate of microbial metabolism. Like the yield equations, they are independent of the type of bioreactor. Unlike the yield equations, rate equations have no theoretical basis and can only be derived from laboratory experiments on the particular waste. Since metabolism depends on so many variables (e.g., concentrations of nutrients and inhibitory products, presence of micro-environments due to adsorption of the cells and/or the contaminants to solid surfaces, pH, temperature), a rigorous rate equation would be impossibly complex. A practical procedure is first to decide which two components are the most critical. For bioremediation, one will be the contaminant, and the selection of the other will depend on economics (which nutrient is most expensive to provide?) or mass-transfer considerations (for aerobic processes, it is usually oxygen). With the help of the yield equations, conditions can be established under which the two selected components control the process rate. Laboratory experiments are then done under these conditions with two component concentrations as variables, all of the other conditions being kept close to those that would exist in the full-scale bioreactor.

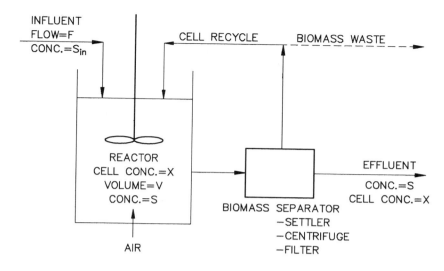

Figure 1.   Schematic for a complete-mix reactor with cell recycle.

These experiments, whose objective is fundamental information about the rate of metabolism that can be applied to *any* full-scale bioreactor configuration, must not be confused with laboratory or pilot scale simulations of a process in which the bioreactor type has already been decided. The approach of deciding, *a priori*, to use a trickling filter/sequencing batch reactor/complete mix activated sludge/fluidized bed bioreactor, etc. and then demonstrating that the choice is feasible is too prevalent in the industry. This approach, being fed by the business needs of companies that specialize in one type of technology, rarely produces an optimum solution.

The type of laboratory bioreactor used for measurement of the rate equations is an important consideration. The mass conservation equations used must be simple enough to allow the inherent kinetics (i.e, the rate equations) to be "backed out" from the experimental data. A steady-state chemostat is usually chosen because its mass conservation equations are algebraic rather than differential.

3.2 Mass Balance Equations

A complete analysis is given here for the simplest and most common process layout, a continuous stirred tank bioreactor (CSTR or chemostat) with cell recycle (Figure 1). The reactor design includes the conventional complete-mix activated sludge system in which the cell separator is a settling tank, as well as various other schemes in which the separator is a centrifuge or membrane system. The disadvantages of this layout, when faced with the challenges from Section 2, are discussed herein and used as a starting point to illustrate improved bioreactor systems. The word "component" is used to include any contaminant, nutrient, or product. Note that the specific rate of consumption, q, will be negative for a product. All concentrations are either molar or, for organic compounds, in terms of carbon equivalents (see Section 3.3).

The mass balance equation for the biomass states that the rate of accumulation equals the rate at which the biomass is growing minus the rate at which it is washed out of the system.

$$V\frac{dX}{dt} = -FX_e + V\mu X \tag{1}$$

The mass balance for a general component in the liquid must also include terms for inflow to the system, gas/liquid mass-transfer, and outflow of the component adsorbed on the biomass. Since adsorption onto single cells or small clumps of cells is a rapid process compared to cell growth, biosorption is described by an equilibrium adsorption isotherm, $A(S)$, that gives the amount adsorbed per unit of cells as a function of the liquid-phase concentration.

$$V\frac{dS}{dt} = F(S_i - S) + Vk_L a(S^* - S) - VqX - FX_e A(S) \tag{2}$$

One of the main benefits of modeling is that it provides exact definitions for the process variables that are independent of scale and important in the design and operation of real systems. The following set can be identified:

- Hydraulic residence time: $\tau = V/F$. Sometimes the hydraulic residence time is expressed as its reciprocal, the dilution rate.
- Mean cell residence time: $M = $ (biomass in system)/(biomass outflow from system). For the system in Figure 1, $M = \tau X/X_e$. In many systems, $X_e = 0$, and cell wastage is provided by a separate bleed from the recycle line or the reactor. This approach changes the details of how M is defined, but not its importance or the form of the final equations. Note that $M \geq \tau$ with the equality applying when there is no cell recycle.
- Loading: $L = $ inflow rate of a component per unit reactor volume $= (k_L a S^* + S_i/\tau)$. The first term is added to generalize the usual definition to components that may be introduced in either the liquid or the gas stream. For components like oxygen and methane, this term will be dominant, and the liquid-phase term is usually ignored even though the dissolved gas concentration in the liquid influent may not be exactly zero.
- Effective inlet concentration: $S_{in} = (S_i + k_L a\tau S^*)/(1+k_L a\tau)$. For a component that may be introduced in either the gas or liquid phase, this is an average of the inlet concentration in the liquid (S_i) and gas (S^*) phases weighted according to the relative importance of the two routes (i.e., the magnitude of $k_L a\tau$).

Substituting these definitions into Equations 1 and 2 gives Equations 3 and 4.

$$\frac{dX}{dt} = X\left(\mu - \frac{1}{M}\right) \tag{3}$$

$$\frac{dS}{dt} = L\left(1 - \frac{S}{S_{in}}\right) - X\left(q + \frac{A(S)}{M}\right) \tag{4}$$

Two points should be noted. First, the hydraulic residence time, τ, mass-transfer coefficient, $k_L a$, and cell separation efficiency, X/X_e, do not appear explicitly in Equations 3 and 4, only combined as L and M. Second, the limiting nutrient can be defined as the nutrient whose concentration (S) reaches zero first as the cell concentration (X) increases. Assuming steady state conditions with no biosorption, it is clear from Equation 4 that the limiting nutrient is the one with the lowest value of L/q, which can be interpreted as supply (L) divided by demand (q). The design procedure to prevent oxygen limitation based on this generalization of the "loading" concept is given in Section 4.

For any volatile component, a mass balance can also be written for the gas phase that relates the inflow rate, the outflow rate, and the amount transferred to or from the liquid phase.

$$(QC)_i - (QC)_e = k_L a\,(S^* - S) \tag{5}$$

Q is the total volumetric gas flow rate per unit volume of liquid (the so-called VVM gas rate). It is different at the inlet and outlet due to changes in temperature and pressure as well as transfer of some gases (O_2) to the liquid and others (CO_2, water vapor) from the liquid. In a mechanically agitated CSTR, the gas phase is taken to be completely mixed, so the saturation concentration, S^*, is the value in equilibrium with the *outlet* gas composition. If Henry's law applies to this component, then:

$$HS^* = C_e \tag{6}$$

The value of $k_L a$ is not the same for different components because it is a function of the component's diffusivity in the liquid phase (D). If the $k_L a$ value for oxygen is known (subscript O), then a reasonable approximation for any other component is:

$$k_L a = (k_L a)_o \left(\frac{D}{D_o}\right)^{2/3} \tag{7}$$

3.3 Yield Equations

The common nutrients and products are listed in Table 1 with the nomenclature to be used in this section. Inorganic compounds are given in molar terms. Organic compounds or compounds of unknown composition are shown as carbon equivalents, i.e., the amount of the compound containing 1 mole of carbon. Rates and concentrations must follow the same conventions. For example, the specific rate of oxygen consumption, q_o, is defined as moles of O_2 consumed per carbon

Table 1.   Components in Yield Equations

		Identifying Subscript	Composition
Nutrients	Contaminant	None	—[a]
	Oxygen	O	O_2
	Auxiliary nutrient	A	—[a]
	Electron acceptor	E	—[a]
Either nutrient or product	Nitrate		HNO_3
	Sulfate		H_2SO_4
	Phosphate		H_3PO_4
	Carbon dioxide		CO_2
	Water		H_2O
Products	Biomass	B	—[a]
	Metabolic product	M	—[a]
	Respiratory product	R	—[a]
	Halide ion	C	HCl

[a]   One carbon equivalent of any organic compound is given the generic formula $CH_hH_nO_oS_sP_pCl_c$.... . Subscripts on h, n, ..., etc., identify which compound they come from.

equivalent of biomass per hour. When converting this to a "dry weight of biomass" basis, a correction is needed for the ash content of the biomass.[2] A concentration in carbon equivalents per liter can readily be measured as total organic carbon divided by 12. When the chemical composition of the waste stream is unknown, the h, n, o, and s values in the empirical formulae shown in Table 1 can be measured by modern elemental analyzers, and total organic halide analyzers are available to give the c value. The "auxiliary nutrient" may be simply ammonia, yeast extract, etc., added to nitrogen-deficient wastes. However, in cometabolic processes, it is the main carbon and energy source for the biomass.

Many biological waste treatment processes including municipal sewage treatment generate sulfate, nitrate, and phosphate because the waste is rich in sulfur-, nitrogen-, and phosphorus-containing compounds. Other wastes deficient in these elements must be supplemented with sulfate, nitrate, and phosphate nutrients. For autotrophic bacteria (e.g., *Nitrosomonas* and *Nitrobacter*) used for waste treatment, CO_2 becomes a nutrient. Nitrate, sulfate, and CO_2 must also appear on the "nutrient" side of the mass balance equations when used as electron acceptors by anaerobic microorganisms.

Table 1 shows three products of microbial metabolism. The first product is the extra biomass produced by microbial growth. One reason for the general utility of the analysis given here is that the elemental composition of biomass is surprisingly constant, independent of the type of microorganisms and the growth substrate (some measured compositions are shown in Table 2). The second category of product is the end product of anaerobic respiratory processes: N_2, H_2S, or CH_4 for denitrifying, sulfate-reducing, or methanogenic processes, respectively. The "metabolic product" in Table 1 covers all other possibilities, e.g., products of incomplete mineralization and end products of fermentative metabolism.

Table 2.   Measured Values of Cell Composition[14]

Organism	Empirical Formula	Mass Fraction of Ash	γ_B
Bacteria	$CH_{1.67}N_{0.2}O_{0.27}$	0.08	6.13
Activated sludge	$CH_{1.4}N_{0.2}O_{0.4}$	0.083	5.60
A. aerogenes	$CH_{1.78}N_{0.24}O_{0.33}$	0.089	6.32
S. cerevisiae	$CH_{1.62}N_{0.15}O_{0.53}$	0.053	5.31
C. utilis grown on:			
Glucose $\mu = 0.08$ h^{-1}	$CH_{1.82}N_{0.19}O_{0.47}$	—	5.83
Glucose $\mu - 0.45$ h^{-1}	$CH_{1.84}N_{0.2}O_{0.56}$	—	5.72
Ethanol $\mu = 0.06$ h^{-1}	$CH_{1.82}N_{0.19}O_{0.55}$	—	5.67
Ethanol $\mu - 0.43$ h^{-1}	$CH_{1.84}N_{0.2}O_{0.55}$	—	5.74

The objective of the yield equations is to relate the rates of consumption and production of all of these nutrients and products to each other. This is done by reducing microbial metabolism to five pseudo-chemical reactions. More reactions must be added if the analysis is generalized to include more nutrients and products. Each pseudo-reaction produces or consumes metabolic energy in the form of adenosine triphosphate (ATP) and reducing equivalents written for convenience as $\hat{H}$ [actually $H^+ + e^-$ and attached to a carrier such as nicotinamide adenine dinucleotide (NAD)]. The yield equations are the resulting intracellular element and energy balances. The basis for this procedure is taken as one carbon equivalent of biomass, so the rates are the specific rates, μ for cell growth and q (with appropriate subscript) for other components. The pseudo-reactions are as follows:

- Pseudo-reaction 1: The contaminant is oxidized at a rate r_1, C-quiv./C-equiv. biomass hour.

$$CH_hN_nO_oS_sP_pCl_c + \varepsilon O_2 \rightarrow w\ H_2O + CO_2 + n\ HNO_3 + sH_2SO_4$$
$$+ pH_3PO_4 + c\ HCl + (\gamma + \varepsilon\gamma_o)\ \hat{H}$$

where $w = 2\varepsilon - 2 + o - 3n - 4s - 4p$
$\gamma = 4 + h + 5n - 2o + 6s + 5p - c$

The εO_2 term in this reaction includes only the nonrespiratory uses of oxygen. In most cases, $\varepsilon = 0$. However for highly reduced contaminants, the first step in the biochemical pathway is often oxidation to an alcohol by a monooxygenase enzyme (e.g., $\varepsilon = 1$ for methane). In the metabolism of aromatic compounds, a dioxygenase enzyme is commonly involved in ring cleavage ($\varepsilon = 1/6$ for benzene).

The γ value is the number of available electrons per C-equivalent (or mole) of the contaminant, a measure of its oxidation/reduction state. There is a similar number for every nutrient and product, so for molecular oxygen $\gamma_0 = -4$. Note that this type of analysis is normally done with NH_3 as the base oxidation state for the nitrogen atom[18] changing the nitrogen contribution in the definition of γ to $-3n$. Nitrate is used here as the base oxidation state because it is the usual product of a well-run wastewater treatment plant and because of its potential importance as an electron acceptor. This approach brings nitrogen into line with carbon, hydrogen, sulfur, and phosphorus for which the most oxidized condition is taken as the base state (i.e.,

$\gamma = 0$ for HNO_3, CO_2, H_2O, H_2SO_4, and H_3PO_4). With this definition, the γ values for the organic fraction of biomass vary between 5.3 and 6.3 (Table 2), so an average value $\gamma_b = 5.8$ will be "close enough for engineering purposes" in many situations.

Each time this pseudo-reaction happens, it generates a fixed number of moles of ATP, N, a value that can only be found from detailed knowledge of the biochemistry involved. For example, $N = 1/3$ mol/C-equiv. for the catabolism of glucose via the Emden-Meyerhof-Parnus (EMP) pathway and the Krebs cycle, and $N = 0$ for methane.

- Pseudo-reaction 2: The auxiliary nutrient may be broken down in the same way generating reducing power at a rate $r_2(\gamma_A + \gamma_o\varepsilon_A)$ and ATP at a rate r_2N_A. For cometabolic processes, this pseudo-reaction and subsequent respiration are the main source of energy for the organisms. The details and discussion are the same as for Reaction 1.

- Pseudo-reaction 3: The anabolic pathways build up new biomass from the contaminant and/or the auxiliary nutrient at a rate equal to the specific growth rate μ C-equiv./C-equiv. hour.

$$\beta CH_hN_nO_oS_sP_pCl_c + \beta_A[CH_hN_nO_oS_sP_p]_A + \beta_oO_2 +$$
$$(n_B - \beta n - \beta_An_A)HNO_3 + (s_B - \beta s - \beta_As_A)H_2SO_4 +$$
$$(p_B - \beta p - \beta_Ap_A)H_3PO_4 + (\gamma_B - \beta\gamma - \beta_A\gamma_A - \beta_o\gamma_o)\hat{H} \rightarrow$$
$$[CH_hN_nO_oS_sP_p]_B + (\beta + \beta_A - 1)CO_2 + \beta_cHCL + \beta_wH_2O$$

The value of β_w can be found, if needed, from an oxygen balance over this reaction leaving three unknowns β, β_A, and β_o, which can only be found from knowledge of biochemical pathways. The oxygen, nitrate, and sulfate appearing in this reaction include only the nonrespiratory uses of these compounds. In some situations, nitrate, sulfate, phosphate, and $\hat{H}$ may be products of this reaction, and CO_2 may be a nutrient. The relevant coefficients in this reaction will then be negative.

Each time the above reaction happens it consumes, by definition, $1/Y_{ATP}$ moles of ATP, where Y_{ATP} is expressed as C-equivalents of biomass produced per mole of ATP consumed. Y_{ATP} is another parameter that can be assumed constant as a first approximation. It is, in fact, slightly larger when the auxiliary nutrient is a rich media like yeast extract than when the cell must manufacture all of its amino acids, nucleotides, etc., from simple precursors such as sugars, ammonia,[19] etc.

- Pseudo-reaction 4: The biomass in Reaction 3 can be replaced by any other metabolic product, the corresponding reaction rate being q_M. The concept of Y_{ATP} can be generalized to Y^M_{ATP}, the number of carbon equivalents or moles of metabolic product that can be made using 1mole of ATP. The β coefficients for this reaction are identified by the superscript M.

- Pseudo-reaction 5: For fermentative processes, the net production of $\hat{H}$ by Reactions 1 through 4 must be zero. Respiratory organisms can convert excess $\hat{H}$ to energy (ATP) by passing the electrons to an external electron acceptor.

$$[CH_hN_nO_oS_s]_E + (\delta\gamma_R - \gamma_E)\hat{H} \rightarrow \delta[CH_hN_nO_oS_s]_R + (o_E - \delta o_R)H_2O$$

If the rate of this reaction is r_5 mole/h, ATP is produced at a rate $r_5(P/O)(\delta\gamma_R - \gamma_E)/2$, where (P/O) is defined as the amount of oxidative phosphorylation per pair of electrons passed down the cytochrome chain.

We can now write the following conservation equations:

Contaminant: q $=$ $r_1 + \beta\mu + \beta^M q_M$
Auxiliary nutrient: q_A $=$ $r_2 + \beta_A\mu + \beta_A{}^M q_M$
Oxygen (if present): q_o $=$ $\varepsilon r_1 + \varepsilon_A r_2 + \beta_o\mu + \beta_o{}^M q_M + r_5$
Respiratory product: q_R $=$ δr_5
(if O_2 not present)   (8)
Reducing equivalents: $r_1(\gamma + \gamma_o\varepsilon) + r_2(\gamma_A + \gamma_o\varepsilon_A) +$
$\mu(\beta\gamma + \beta_A\gamma_A + \beta_o\gamma_o - \gamma_B) +$
$q_M(\beta^M\gamma + \beta_A{}^M\gamma_A + \beta_o{}^M\gamma_o - \gamma_M) =$
$r_5(\delta\gamma_R - \gamma_E)$

$$\text{ATP}: \quad r_1N + r_2N_A + r_5(\delta\gamma_R - \gamma_E)\frac{(P/O)}{2} = \frac{\mu}{Y_{ATP}} + \frac{q_m}{Y_{ATP}^M} + k$$

The k in the final equation is the amount of ATP needed per hour for the maintenance of 1 C-equiv. of biomass. Eliminating the unknown rates r_1, r_2, and r_5 gives the two basic yield equations.

$$\sum_{\text{substrates}} \gamma q = \sum_{\text{products}} \gamma q \tag{9}$$

This first equation is clearly an oxidation/reduction balance for microbial metabolism. The "substrates" term includes the contaminant, auxiliary nutrients, and oxygen (if present). The anaerobic electron acceptors (HNO_3, H_2SO_4, CO_2) make no contribution to this sum because their γ values are zero. The "products" term includes microbial growth (μ replaces q) and any products of anaerobic respiration as well as metabolic products. The second equation represents an intracellular energy balance.

$$\alpha q + \alpha_A q_A = \alpha_B\mu + \alpha_M q_M + k \tag{10}$$

The α parameters represent the amount of ATP associated with the consumption or production of 1 mole or carbon equivalent of a component. Thus for substrates, $\alpha = N + (\gamma + \gamma_o\varepsilon)(P/O)/2 =$ maximum ATP available from complete catabolism via both substrate-level and oxidative phosphorylation. For products, $\alpha_B = (1/Y_{ATP}) + \beta N + \beta_A N_A + [\gamma_B + \gamma_o(\beta\varepsilon + \beta_A\varepsilon_A - \beta_o)]$ (P/O)/2 = total ATP cost of making 1 C-equiv. of biomass. The first term in this definition is the amount of ATP consumed to make biomass, while the remaining terms are the amount of ATP production foregone by not using the substrates for energy production. There is a similar definition for the metabolic product α_M. Note that when metabolism involves an oxidase enzyme, it usually appears early in the pathway and, therefore, appears in both the catabolic (1 or 2) and the anabolic (3 or 4) reaction pathways. Then $\beta_o = \beta\epsilon + \beta_A\epsilon_A$, and the definitions of α are simplified.

Equation 10 is identical to the "Linear Growth Equation" demonstrated experimentally by Stouthammer and Verseveld[20] for many types of bioprocesses. The advantage of the theoretical derivation given here is that it relates the parameters, α, to biochemical constants such as Y_{ATP} and (P/O) whose values are known approximately. Several known results can be derived immediately from these equations. Consider, for example, the complete aerobic mineralization of a contaminant that is the sole carbon and energy source for metabolism. The yield equations would conventionally be written:

$$q = \frac{\mu}{Y} + k_m \qquad \text{(a)}$$

$$q_o = \frac{\mu}{Y_o} + k_o \qquad \text{(b)}$$

(11)

The yield (Y, Y_o) and maintenance coefficients (k_m, k_o) are usually be treated purely as empirical constants. However, comparing Equation 11 with the suitably simplified version of Equations 9 and 10 and setting $\gamma_o = -4$ give the relationships $k_m = k/\alpha$, $k_o = k_m\gamma/4$, $Y_o = 4Y/(\gamma - \gamma_B Y)$ in which many of the parameters $(k, \gamma, \gamma_B, \alpha)$ are known or can be estimated. Also, if the generation of ATP by substrate-level phosphorylation is small compared with oxidative phosphorylation then:

$$Y = \frac{\alpha}{\alpha_B} = \frac{(\gamma - 4\epsilon)(P/O)}{\left[\gamma_B - 4(\beta\epsilon - \beta_A\epsilon_A - \beta_o)\right](P/O) + 2/Y_{ATP}} \qquad (12)$$

It follows that the cell yield coefficient is proportional to the effective oxidation/reduction state of the contaminant $(\gamma - 4\epsilon)$ or, equivalently, that the "yield on available electrons" or the "COD yield" is a constant. This has been observed in many experiments and Equation 12 with suitable values of (P/O) and Y_{ATP} and after correction for maintenance gives values that agree with the observed yields. Note that the "4ϵ" correction for nonrespiratory oxygen consumption is included here for the first time and corrects a discrepancy that existed previously between this analysis and observed cell yields for methanotrophic bacteria.[2]

There are several important nutrients and products (Table 1) for which $\gamma = 0$ and which consequently do not appear in the yield Equations 9 and 10. The net rates of production or consumption of CO_2, HNO_3, H_2SO_4, etc., can, if needed, be related to the "known" rates (q, q_A, μ, q_M) by taking element balances on carbon, nitrogen, sulfur, etc., over the set of reactions discussed previously. For example, a process carried out by denitrifying bacteria in the absence of molecular oxygen would produce a nitrate balance:

$$q_N = r_5 + (n_M - \beta^M n - \beta_A^M n_A)q_M + (n_B - \beta n - \beta_A n_A)\mu - nr_1 - n_A r_2$$

Setting $\delta = 1/2$ (since R is N_2) and substituting from Equation 9 gives the nitrogen balance:

$$q_N = 2q_R + n_B\mu + n_M q_M - nq - n_A q_A \qquad (13)$$

Once any three of the rates q, q_A, q_M, q_R, μ, or q_N are known, the other three can be calculated from Equations 9, 10, and 13.

3.4 Rate Equations

The mass balance and yield equations have a solid theoretical foundation in the conservation of mass and energy over the bioreactor and the biomass respectively (the energy balance over the bioreactor is not discussed here, but fixes the temperature of operation). The resulting equations can, therefore, be used with confidence for preliminary process design even if the only experimental data available is the identity of the main nutrients and products. The rate equations are completely different in this respect. They attempt to describe, in simple mathematical form, the incredibly complex processes that determine the speed of microbial metabolism, not a task that can be approached with confidence. Rate equations can only be found by experimentation, and before this is attempted, three preliminary questions must be answered.

3.4.1 What Are the Limiting Nutrients?

The limiting nutrient is the one that stops the desired microbial metabolism by being consumed first. Since the objective of waste treatment is to remove the contaminant, one task of the process designer is to ensure that the contaminant is the limiting nutrient. This can be accomplished by nutrient addition and appropriate process configuration.

For aerobic processes, it is convenient to think of two limiting nutrients: one in the gas phase and one in the liquid. One of these is the contaminant, which may be in the gas in the general analysis given here (see Section 4.6). The other is usually decided by economic considerations. Adding nutrients to a process costs money, and common sense requires that the most expensive be used most sparingly. The gas-phase limiting nutrient is particularly important because the process will operate in the mass-transfer limited regime for this nutrient, and it is this mass-transfer capacity that ultimately fixes the size of the bioreactor.

Some examples will clarify this discussion. For the aerobic treatment of an organic wastewater deficient in nitrogen, process designers calculate the required addition of ammonia (i.e., the ammonia loading) using the L/q criteria given by Equation 4 coupled with the yield equations. Since ammonia is expensive, they also consider reducing the amount needed by running the process at a high mean cell residence time so that more contaminant is used for maintenance than growth (see Section 4.1). The gas-phase limiting component is oxygen, and L/q for oxygen must be slightly larger than L/q for the contaminant, or the bioreactor will become overloaded and anaerobic. This requirement fixes the contaminant loading and thus the bioreactor volume for a given flow of waste.

For the degradation of dissolved halocarbons by methanotrophic bacteria, the contaminant must be made the limiting component in the liquid by appropriate addition of mineral salts. The gas phase contains both methane and oxygen. Due to its cost, methane should be made limiting, while the oxygen in air is free. The composition of the gas phase needed to accomplish this can be calculated from the L/q criteria, the yield equations, and Equation 7.

Anaerobic processes usually have no gas-phase limiting nutrient, although it may be necessary to strip products such as H_2S out of the liquid before they become inhibitory to the process either directly or through their effect on pH. Since there are no gas/liquid mass-transfer limitations on the reaction rate, active biomass concentrations can be extremely high. Modern anaerobic processes such as the Upflow Anaerobic Sludge Blanket Reactor[1] have very high productivities despite the slow inherent kinetics of the microorganisms.

3.4.2 How Many Rate Equations Are Needed?

The systematic approach to the yield equations given in the previous section fixes the number of degrees of freedom of the system and thus the number of rate equations needed. Consider the simplest case in which the contaminant provides the carbon and energy source for an aerobic microorganism while being completely mineralized. The yield equations reduce to Equations 11(a) and 11(b), two equations for three unknown rates (μ, q, q_o) leaving one degree of freedom and thus one rate equation that must be specified. The same applies to anaerobic processes that generate a single product besides CO_2 (anaerobic digestion to methane for example). The oxygen consumption rate, q_o, is replaced by a product formation rate (q_M for a fermentative process, q_R for a respiratory one), and there is still one degree of freedom. If significant amounts of a second product are generated (e.g., H_2 in anaerobic digestion), then there are four unknown rates, and two rate equations are needed. The same applies to cometabolic degradation by respiratory organisms. The four unknowns are q (contaminant), q_o (oxidant), μ (growth), and q_A (carbon and energy source); any two of which must be specified by rate equations. Each additional auxiliary nutrient and product adds one extra degree of freedom.

3.4.3 Which Rates Should Be Specified?

All of the rates discussed above (q, μ, q_o, ...) depend on the concentrations of the limiting nutrient(s) to which the cells are exposed. Which one or two should be described by means of rate equations, and which should be left to be calculated from the yield equations? The most common choice is to describe the specific growth rate of the biomass as a product of Monod-type functions.

$$\mu = \hat{\mu} \; \frac{S}{K+S} \; \frac{S_g}{K_g + S_g} \tag{14}$$

The subscript g here describes the gas-phase limiting nutrient. When this is oxygen, the half-velocity constant (K_g) is small, and the Monod function can be replaced by the Heaviside function $H(S_g)$, which equals 1 when $S_g > 0$ and 0 when $S_g = 0$. This part of the equation is sometimes left out completely, which is acceptable as long as the actual bioreactor is never overloaded (i.e., $S_g \gg K_g$ at all times).

Even if the algebraic formulation of Equation 14 is correct (other forms have been proposed, but they are difficult to distinguish experimentally), there is an obvious problem in the common case where the limiting nutrient is the carbon and energy source. The net specific growth rate has been defined here as μ. However, it is known that microbes within a zero carbon-source environment ($S = 0$) do not only stop growing ($\mu = 0$) as suggested by this equation, but actually start to decay, die, and lyse ($\mu < 0$). Even worse, putting $S = \mu = 0$ in Equation 11(a) gives $q = k_m$, which implies that uptake of the carbon source continues even when none is present. This is illogical and can lead to absurd predictions including negative concentrations. The solution, first realized by Lawrence and McCarty,[21] is to specify, not μ, but q by Monod-type functions. Now $q = 0$, $\mu = -Yk_m$ [Equation 11(a)], and $q_o = Yk_m\gamma_B/4$ [Equation 11(b)] when $S = 0$, which are at least qualitatively correct. The checking of the limits of the rate equations is very important and often requires considerable judgment. For example, formulating the rate equations such that $q = 0$ when the dissolved oxygen $S_o = 0$ ignores the possibility of facultative anaerobic activity in the process and may also produce the absurd prediction that q_o is positive when $S_o = 0$. These problems can only be resolved on a case by case basis.

Since many of the hazardous contaminants considered here are inhibitory at high concentrations, the substrate-inhibition rate equation is a useful generalization of the Monod equation.

$$q = \frac{\bar{q}S}{S + K\left[1 + \left(S/S_m\right)^2\right]} \tag{15}$$

The equation is written in this nonstandard form in order to introduce explicitly the parameter S_m, which is the concentration at which the specific substrate uptake rate, q, is a maximum (see Figure 2). Note that this maximum value is not $\bar{q}$, but $\bar{q}/(1 + 2K/S_m)$.

4. CONSEQUENCES FOR PROCESS DESIGN

The mass balance, yield, and rate equations described in the previous section together constitute a complete process model: the mass conservation equations

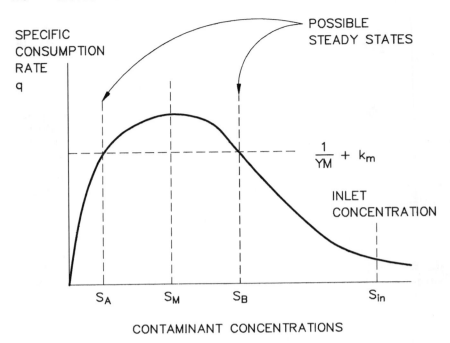

Figure 2.   Steady states in a continuously stirred tank reactor with substrate inhibition.

describing the reactor configuration, while the yield and rate equations describe the inherent microbial metabolism. A study of this model allows many of the process variables to be fixed in terms of the metabolic parameters. Consider first the case of a soluble, contaminant that does not biosorb, but can be mineralized by aerobic bacteria and is inhibitory at high concentrations (e.g., phenol).

4.1 Mean Cell Residence Time

At steady-state, Equation 3 reduces to $\mu = 1/M$. If the contaminant is the limiting nutrient (i.e., excess N and P are provided, and O_2 is transferred at a sufficient rate), this can be substituted in Equations 4 and 11(a) to give:

$$q = \frac{1}{YM} + k_m = \frac{\overline{q}S}{S + K\left[1 + \left(S/S_M\right)^2\right]} \tag{16}$$

$$Xq = L\left(1 - \frac{S}{S_{in}}\right) \tag{17}$$

Equation 16 is illustrated graphically in Figure 2. It is clear that there are two possible steady-state effluent concentrations, S_A and S_B, one above and one below the concentration, S_M, at which the rate of microbial metabolism is a maximum. These correspond to the two solutions of the quadratic Equation 16. The solution S_B obviously exists only if S_B is less than the inlet concentration, and even then, it can be shown that this steady-state is unstable. It is significant only during start-up. If the flow through the system is turned on while the contaminant concentration in the reactor is larger than S_B, then the inoculum of biomass will be washed out. If $S < S_B$, then the system may (if M is large enough) approach the stable-steady state at S_A.

Since the system always operates below S_M, it follows that a chemostat, with or without cell recycle, is not a good experimental system in which to study substrate inhibition. The behavior of the metabolic rate curve in Figure 2 for $S > S_M$ cannot be studied in a chemostat. The insensitivity of the system to substrate inhibition is further illustrated in Figure 3, which gives the results of the analysis in dimensionless variables for $S_M \gg K$ (i.e., no substrate inhibition) and $S_M = 5K$, a typical value for inhibitory substrates. There is little practical difference between the sets of curves, except near the washout limit (small M) where the system would not normally be operated.

The most surprising result of the above analysis is that the steady-state effluent contaminant concentration depends only on M and the parameters $(Y, k_m, K, \bar{q})$ that describe microbial metabolism. It is independent of both the influent concentration and the hydraulic residence time. The reason is shown by Figure 3. Changing the inlet concentration in the range $5K < S_{in} < \infty$ makes little difference to the dimensionless cell concentration $X\bar{q}/L$ (going down to $S_{in} = K$ would make a difference; very dilute wastes require special design procedures). It follows that the cell concentration in the system, X, is proportional to the loading, L. Whenever L is increased, either by increased wastewater concentration or flow rate, the process responds automatically by producing more biomass and keeping the effluent concentration at the same level as long as the process control system keeps M constant.

The lower limit of M is the biomass washout point.

$$\frac{YM\bar{q}}{1 + YMk_m} = \text{the larger of} \quad \begin{cases} 1 + 2K/S_m \\[2mm] 1 + \dfrac{K}{S_{in}}\left[1 + \left(\dfrac{S_{in}}{S_M}\right)^2\right] \end{cases} \tag{18}$$

The upper limit $M = \infty$ corresponds to complete cell recycle with no biomass wastage. However, even at $M = \infty$, the effluent concentration is not zero. What happens as less and less biomass is wasted is that the biomass concentration increases until it reaches the value $X = L/k_m$ at which all incoming contaminant must be used for maintenance and none is left for growth. The quantity on the abscissa of Figure 4 approaches $\bar{q}/k_m$, and the corresponding minimum value of S is $Kk_m/(\bar{q} - k_m)$. (Note that since no growth is occurring, the value of Y is irrelevant.) This minimum S value is negligibly small for aerobic processes where

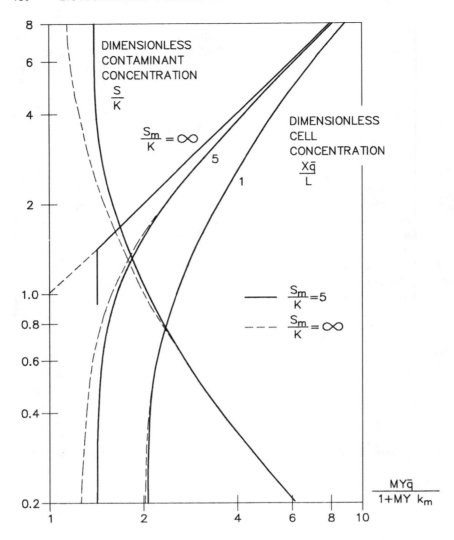

Figure 3. Relationship between the mean cell residence time, M, the cell concentration, X, and the liquid-phase contaminant concentration, S, in a complete-mix system.

$\bar{q}$ is typically 100 times k_m, but it may be significant for the less efficient types of anaerobic metabolism, which have a higher $k_m/\bar{q}$ ratio. This is one basis for the conventional belief that anaerobic digestion cannot produce high quality wastewater treatment effluents (also see Section 4.4).

Operating at the M = ∞ limit is unwise because cell death is inevitable and, with no biomass outflow, the reactor will eventually fill with dead cells. Operating at the point where equal amounts of contaminant are used for growth and maintenance (M = 1/Yk_m) makes little practical difference to the effluent concentration and avoids this problem. It does, however, increase the amount of biomass for

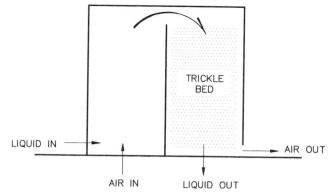

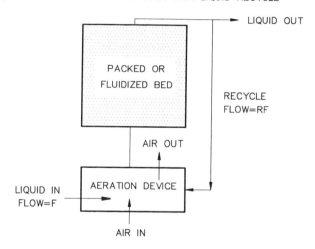

Figure 4. Processes for combining mixed and plug-flow sections. (A) Bubble column plus trickling filter and (B) an immobilized cell bioreactor with liquid recycle.

disposal. Note that the "extended aeration" activated sludge system, operated with no deliberate cell wastage, is self-controlling in this sense. M is not actually infinite because the biomass tends to deflocculate and leaves the system in the overflow of the secondary clarifier.

4.2 Loading and Reactor Size

One reason for the popularity of complete-mix activated sludge systems is that experience confirms the results of the above analysis. If M is kept constant, the effluent contaminant concentration remains low, constant, and virtually indepen-dent of the inlet flow and concentration. However, there must obviously be some upper limit to the loading. What "overloading" means in these systems is that the aeration system cannot keep up with the demands of contaminant degradation. The gas-phase limiting nutrient (oxygen) becomes more critical than the liquid-phase limiting nutrient (the contaminant), and the bioreactor becomes anoxic. The analysis in the previous section, based on the contaminant being the critical nutrient, breaks down.

It was shown previously that the limiting nutrient is the one with the smallest value of L/q. Substituting from the yield equations and the general definition of loading used here (ignoring the oxygen dissolved in the influent), it follows that the criteria for oxygen limitation is:

$$L > \frac{q}{q_o}\left(k_L aS^*\right)_o = \frac{4\left(k_L aS^*\right)_o}{\gamma - \gamma_B Y/\left(1 + Yk_m M\right)} \tag{19}$$

The actual loading should be set at a value slightly below this. The loading and the influent wastewater concentration fixes the hydraulic residence time and thus the size of the reactor for a given flow. This completes the reactor design problem; given the characteristics of the wastewater (F, S_{in}), the bacterial culture ($\bar{q}$, S_m, Y, k_m, K), and the reactor aeration system ($k_L aS^*)_o$, we have specified the reactor volume and the efficiency of the cell separation device $X/X_e = M/\tau$ (or equivalently, the cell wastage rate) needed to achieve a given effluent quality.

For very concentrated wastewaters, this calculation may give $M < \tau$, in which case the cell separator makes little sense. The correct approach is then to dilute the influent, either with recycled effluent or with a separate aqueous nutrient stream, until the calculation produces $M = \tau$, which means that cell recycle is not needed. The limit of this process for pure nonaqueous waste streams is essentially a fed-batch bioreactor. The waste, plus an aqueous nutrient stream of precisely calculated flow and concentration, are pumped slowly into a large, aerated agitated tank. Some liquid from the tank is wasted periodically to keep τ (=M) at the required value.

4.3 Substrate Inhibition and Axial Mixing

Having completed the design of the system in Figure 1, there remains the question of whether this reactor configuration is in fact best (i.e., most economical for the given task). This is not a question that can be answered definitively, since it is impossible to repeat the technical and economic calculations for all alternative process flowsheets in a reasonable time. Only general guidelines can be given. The first concerns whether mixing is a good idea or whether a plug-flow type of reactor, a packed-bed immobilized cell system for example,[8] may perform better.

The rule in chemical reaction engineering is that if the reaction rate decreases as the reaction proceeds (positive order kinetics, the usual situation), then a plug-flow reactor will be more productive. The biodegradation of an inhibitory substrate is a rare practical example of negative order kinetics. As the reaction proceeds, biomass is produced, and substrate concentration is reduced, both of which tend to increase the reaction rate. The complete-mix reactor is therefore optimal, at least for reducing the concentration to a value close to S_M (see Figure 3).

If the objective is to reduce S to a value much smaller than S_M, then the situation is less clear. Reducing S will decrease the specific consumption rate. The theoretical optimum solution is a continuous stirred tank bioreactor (CSTR) operating just below S_M, followed by a plug-flow reactor. This can be approximated in several ways. One is to add a trickle-bed section to the overflow of a CSTR as shown in Figure 4(A). Mechanical agitation of the first section is optional. As the liquid overflow trickles through the second section, the contaminant concentration is decreased to the desired effluent value by the bacteria present. Oxygen is provided by the air leaving the CSTR.

A simpler solution is the packed or fluidized bed, immobilized cell bioreactor with liquid recycle as shown in Figure 4(B). The recycle has the same effect as mixing, and for large recycle ratio R, this process is functionally identical to that of Figure 1, except that a high mean cell residence time is maintained not by a cell separation device, but by immobilizing the biomass in gel beads, porous particles, or as a biofilm on a solid support. For smaller R values (<10 approximately), there are definite concentration gradients through the bed, and it can be thought of as a plug-flow bioreactor. If the waste concentration $S_{in} \gg S_M$, then, in the absence of recycle, the biomass near the reactor entrance would be exposed to high contaminant concentrations and, thus, completely inhibited. However, knowing S_{in} and S_M, it is possible to specify the recycle ratio R such that the concentration at the bed inlet is slightly larger than S_M. The concentration is then decreased to the required effluent value in the plug-flow bed by biomass that is completely uninhibited.

When these systems are used for aerobic processes, sufficient oxygen must be dissolved in the bed influent to prevent anaerobic conditions from developing. In practice, this represents another constraint on R, and an optimum value must be found that minimizes substrate inhibition while allowing sufficient aeration. The Envirex system[22] for aerobic hydrocarbon degradation is an excellent example of this type of process. It uses a fluidized bed reactor system, which gives a much larger surface area of biofilm than can be achieved in a packed bed. The biofilm is supported on granular activated carbon, which protects against process upsets due to shock loads, since the excess contaminant is adsorbed on the activated carbon and then slowly released for biodegradation. A pure-oxygen aeration system is used, which means that less compromise must be made on the recycle ratio to achieve good aeration. Note that with fluidized beds, it is also possible to provide oxygen by bubbling air directly through the bed.[23] However, the aeration induces so much axial mixing that the system is almost functionally identical to Figure 1, and whatever advantage may result from having a plug-flow section is lost.

4.4 Limitations on Bioreactor Productivity

It was shown in Section 4.2 that for the aerobic treatment of a reasonably concentrated waste, the bioreactor volume (strictly the liquid volume it contains) is fixed by oxygen transfer considerations. This point is so important that it is worth repeating the argument in words rather than in mathematics. The critical feature of bioprocess kinetics is that they are autocatalytic. The microbes not only catalyze the degradation of the waste, but reproduce themselves as part of the process. By retaining these growing cells in the process by cell immobilization or recycle, the cell concentration and thus the potential degradation rate per unit volume can be increased to almost any desired level. The upper limit is imposed not by biology but by the rate at which oxygen can be transferred from the air to the microbes.

There are obviously two situations in which this limit does not apply. The first is anaerobic processes including anaerobic digestion and processes involving soluble electron acceptors such as nitrate and sulfate. In these cases, there is no theoretical upper limit to the cell concentration and loading that can be achieved, except for the physical limit of how many microbes can fit in unit volume of liquid while still allowing flow and mixing. The Upflow Anaerobic Sludge Blanket Reactor[1] is a good example of a system operating close to the cell loading limit. The main practical problems are the long start-up period (due to the slow growth of the organisms involved) and the problems of controlling the pH, nutrient balance, etc., at the very high loadings involved. A fluidized bed consisting of small support particles coated with biofilm can also approach the loading limit because the constraints on biofilm thickness imposed by oxygen diffusion through the film are no longer present.[4] Walker et al.[12] have shown the high loadings can be achieved with these systems designed for the biological removal of nitrate from wastewaters.

A second situation in which oxygen transfer is irrelevant involves wastewaters so dilute that the oxygen normally dissolved in them (up to 8 mg/L, assumed negligible in deriving Equation 19) is sufficient for the process. Little development work has been done on bioreactors for this concentration range, even though there are many potential applications. In the pump-and-treat bioremediation of groundwater, the objective is often to reduce a few parts per million of a contaminant down to a drinking-water standard of a few parts per billion. The previous analysis gives the main characteristics for the bioreactor. Achieving an effluent concentration $S \ll K$ requires a very high mean cell residence time (Figure 3). With extremely low levels of contaminants available for cell growth, the only way to maintain the high biomass concentrations needed for a productive reactor is to minimize cell loss from the reactor. Also, an extended start-up period is needed to develop a high, steady-state biomass concentration. Even with $M = \infty$, the analysis of the complete-mix system predicts a definite lower limit $S = Kk_m/(\bar{q} - k_m)$ to the concentration that can be achieved. This prediction of a minimum concentration (of order 10 ppb for many contaminant/microbe combinations) is confirmed by experience and is sometimes used as an argument against the application of bioremediation to very dilute wastes. However, it must not be forgotten that the prediction was made for the complete-mix system of Figure 1 and may not apply to other bioreactor types.

At these very low concentrations, kinetics approach first-order behavior, and a plug-flow immobilized cell bioreactor has definite advantages. A possible candidate is the simple rapid sand filter. Large populations of nitrifying bacteria are known to develop, after a long "ripening" period, in filters treating groundwater containing low levels of ammonia.[24] Other populations may appear naturally in response to the presence of other contaminants, or specialized populations of cometabolic degraders could be grown by the addition of appropriate nutrients (e.g., methane for the growth of methanotrophic bacteria capable of degrading halocarbon contaminants). Very few bacteria leak through to the filter effluent, so the mean cell residence time is controlled by the frequency of filter backwashing. It is instructive to compare this filter/bioreactor with a conventional trickling filter. The solid particles in the filter/bioreactor can be much smaller than these in the trickling filter for two reasons. First, at very low nutrient concentration, the biofilm cannot grow thick enough to clog the bed, and second, since no aeration is needed, there is no need for air space between the particles. These smaller particles mean a much larger surface area to be colonized by bacteria and thus a bioreactor far better adapted to the low-concentration regime.

4.5 Volatile Contaminants in Wastewaters

Many contaminants of concern including hydrocarbons and halocarbon solvents are volatile. Bioprocess design for the aerobic treatment of wastewaters that contain volatile contaminants is influenced by the need to maximize biodegradation and minimize the extent to which aeration transfers the contaminant to the atmosphere. The complete-mix system of Figure 1 has a definite advantage in this regard.[5] The aerated tank contains the lowest (effluent) contaminant concentration in the system, which automatically minimizes volatilization. But there are a number of other possibilities for reducing air-stripping in this system. With $C_i = 0$, combining Equations 5 and 6 gives the rate of stripping per unit reactor volume as:

$$(QC)_e = \frac{k_L aS}{1 + k_L a/Q_e H} \tag{20}$$

The outcome depends on the value of the dimensionless number ($k_L a/QH$). The mass-transfer factor, $k_L a$, for the contaminant is related to $k_L a$ for oxygen (Equation 7) and is fixed by the need to transfer sufficient oxygen for biodegradation. If the airflow is small and the contaminant is not very volatile (small H), then $k_L a \gg QH$ and the amount of contaminant stripped $(QC)_e$ is minimized, even though its concentration in the effluent gas, C_e, is at its maximum possible value HS (i.e., in equilibrium with the liquid-phase concentration). If the contaminant is volatile and airflow is high, then $k_L a \ll HQ$, the stripping rate, approaches its maximum value, $k_L aS$, although the concentration of contaminant in the effluent air decreases. Clearly the design choices are fixed by the regulatory environment; regulations that specify a maximum value of C_e can best be met by increasing the airflow.

In a more sensible regulatory environment that fixes $(QC)_e$ the kg per day of contaminant that can be volatilized, bubble-aerated reactors are at a disadvantage.

Reducing the airflow means fewer bubbles, less interfacial area, and a lower oxygen transfer rate. Experimental correlations have the form $k_L a = bQ^n$, where b and n are constants that depend on power input (for mechanically agitated systems) and vessel geometry (for bubble-agitated systems). $k_L a/QH$ is now fixed by the volatility of the contaminant and the need to transfer oxygen. Furthermore, the exponent n decreases with scale from close to 0.8 for laboratory reactors to approximately 0.5 at commercial scale, so the air stripping of volatile contaminants tends to get worse with increasing scale. If bubble-aerated reactors must be used with volatile contaminants, they should employ good aeration practice (fine bubble dispersion, tall bubble columns, etc.) to maximize the ratio $k_L a/Q$. Feedback control of dissolved oxygen should be considered to keep airflow to the minimum required by the process. Co-current flow of liquid and air through a number of reactors in series will also reduce volatilization.

Although Equation 20 does not apply directly to trickle bed reactors (the liquid phase in these reactors is not completely mixed), the dimensionless number $k_L a/QH$ must still be an important parameter. The advantage of trickle bed systems is that the gas/liquid interfacial area, a, is fixed by the type of packing material making $k_L a$ almost independent of the gas flow rate. Q can, therefore, be reduced down towards the limit where most of the oxygen in the air is consumed. Co-current flow of air and liquid is essential so that air leaving the process is close to equilibrium with the dilute effluent liquid, rather than the concentrated influent. The process shown in Figure 4(A) will give minimal contaminant stripping due to the co-current flow of air, contaminant vapor, and liquid overflow containing bacteria through the trickle bed section. The contaminant stripped in the bubble column section is transferred from the gas to the liquid as bacteria consume the remaining dissolved contaminant. A biofilm established in the trickle bed would enhance this activity, but is not absolutely essential since it is continuously inoculated from the bubble column overflow.

The use of pure oxygen effectively reduces Q_e while increasing oxygen transfer rates, but aeration systems must be modified so that a high proportion of the oxygen, which is very expensive, is actually transferred and used by the bacteria. In the limit, all of the oxygen is used, $Q_e = 0$, and there is no stripping of contaminant. Pure oxygen activated sludge systems seek to approach this limit by using covered, multi-compartment aeration tanks in which the gas and liquid move in a co-current mode. The Envirex system,[22] similar to that shown in Figure 4(B), uses a sealed aeration device. The pressure is maintained with pure oxygen, and the gas space it contains needs only to be purged periodically to prevent excessive accumulation of N_2, CO_2, etc., that desorb from the liquid. Venting of volatile contaminants from the system is negligible, particularly when only the recycle stream (where S is a minimum) is aerated and not the entire reactor influent as shown in Figure 4(B).

4.6 Gas Treatment Bioreactors

Biological treatment can also be used to remove the vapors of volatile contaminants from gas streams, e.g., hydrocarbons in the air resulting from soil venting operations.[25] The reactors for these processes are sometimes referred to as "gas-phase bioreactors", but this is a misnomer. In virtually all cases, the actual degradation still happens in an aqueous phase inside the microbial cell, and the problems of gas/liquid mass-transfer followed by microbial consumption are similar in principle to those for gaseous nutrients, including oxygen, in normal bioreactors. The differences in process configuration arise from the difference in objective. In aeration, the objective is simply to maximize the oxygen transfer rate per unit reactor volume ($k_L aS^*$, see Equation 19). Since oxygen in the air is free and not an environmental contaminant, the fraction of the oxygen entering the bioreactor that is transferred to the liquid phase and consumed by the microorganisms is not a concern. The actual removal in most bioreactors is 2 to 3% (contrast the case of pure oxygen, which is expensive and must therefore be used more efficiently). Aeration of bioreactors is fundamentally different from the situation in bioreactors used to treat gas streams. The fraction of the contaminant removed is now the main objective, and the reactor must be made large enough to achieve a figure of 90 to 100%.

In the analysis given in Section 3, the concepts of "loading" and "inlet concentration" were generalized to include components in both the gas and liquid phases. The same analysis can, therefore, be applied to the design and operation of gas treatment bioreactors. The first task is to identify the limiting nutrient. Suppose the gas stream contains oxygen (subscript o) and a contaminant that serves as the carbon and energy source (no subscript) both of which must be transferred, at different rates, to the aqueous phase before being metabolized. Since the limiting nutrient is the one with the lower value of L/q, oxygen will be limiting (see Equations 7, 9, 10, and 11).

$$\frac{C_{oi}}{C_i}\left(\frac{D_o}{D}\right)^{2/3} < \frac{q_o}{q} = \frac{4}{\gamma - \gamma_B Y/\left(1 + Yk_m M\right)} \tag{21}$$

If this is the case, then the gas must be diluted with air until the contaminant is limiting. Similar relationships exist to fix the necessary concentrations of any other gas-phase nutrients (ammonia?). The necessary loading of liquid-phase nutrients is fixed by a similar combination of the "low L/q" criteria and the yield equations. The liquid flow into the system needs only to be sufficient to bring in nonvolatile nutrients, wash out any nonvolatile products that may inhibit metabolism (e.g., Cl^- from the degradation of halocarbon solvents), and replace losses from evaporation. This flow will usually be small.

Figure 3 still describes the relationship between the mean cell residence time, M, the cell concentration, X, and the liquid-phase contaminant concentration, S, in a complete-mix system. S can be kept small [down to $Kk_m/(\bar{q} - k_m)$] by making M large, which is easily accomplished by a combination of cell immobilization and the small liquid flow. Thus, in almost all cases, it is possible to keep $S \ll C_i/H$ (the concentration in equilibrium with the inlet gas composition), which is the necessary condition for the bioreactor to be running under mass-transfer limitation. The fractional removal of contaminant can then be found solely from Equations 5 and 6.

$$1 - \frac{C_e}{C_i} = \frac{B(1 - HS/C_i) + 1 - Q_i/Q_e}{B+1} \qquad (22)$$

If $HS \ll C_i$ (mass transfer limitation) and $Q_e = Q_i$ (which will be true for dilute contaminants), then the fractional removal is a unique function of the mass-transfer efficiency number $B = k_L a/HQ_e$. All of the comments from the previous section about the factors determining this number apply equally here. Bubble-aerated systems are not recommended because reducing Q reduces $k_L a$ and thus has little effect on B. The value $B = 19$ required to achieve 95% contaminant removal may simply not be achievable. Trickle bed or spray bioreactors are preferable because $k_L a$ is independent of Q and any value of B is possible.

In gas treatment bioreactors, concerns about axial mixing mainly involve mixing of the gas phase. All of the above applies to the mechanically agitated tank shown in Figure 1, where the gas phase is completely mixed and the equilibrium concentration, S^*, is given by Equation 6. This is far from the optimum situation. As the effluent becomes cleaner (C_e decreases), S^* is reduced in proportion and so is the driving force for mass-transfer ($S^* - S$). In a bubble column, a trickle bed, or the combined system of Figure 4(A), the gas moves in a plug-flow mode, and the equilibrium concentration, S^*, is determined, not by the outlet gas composition, but by the log-mean composition.

$$(S^* - S) = \frac{C_i - C_e}{H \ln\left[\dfrac{C_i - HS}{C_e - HS}\right]} \qquad (23)$$

Equation 22 is now replaced by:

$$1 - \frac{C_e}{C_i} = \left(1 - \frac{HS}{C_i}\right)(1 - e^{-B}) \qquad (24)$$

Now if the system is mass-transfer limited ($S \ll H/C_i$), achieving 95% removal requires $B = 3$, a considerable improvement over $B = 19$ needed by the complete-mix system. The huge advantage of the plug-flow systems is not generally appreciated because it only appears when high fractional removal is a requirement. A 3% removal, a typical number for O_2 removal from air in a fermentor, can be achieved with $B = 0.051$ in a bubble column (Equation 24) and $B = 0.053$ in a mechanically agitated bioreactor (Equation 22). This small advantage is more than offset by the higher $k_L a$ values generated by mechanical agitation.

It may seem surprising that no parameters describing microbial metabolism appear in Equations 22 or 24, but this is the nature of a mass-transfer limited process. The kinetics of an individual organism are irrelevant as long as sufficient biomass can be accumulated (the large-M requirement) to consume the contaminant as fast as it is transferred from the gas to the liquid. Even gas streams that could strongly inhibit metabolism can be treated because, at steady-state, the microbes in the liquid phase are exposed only to the low concentration, S, not to the equilibrium concentration (C_i/H), which may be lethal. However, the design, start-up, and control of a bioreactor for this type of gas stream would require considerable care. The system shown in Figure 4(A) would be excellent for this application if most of the "liquid out" stream was recycled to the bubble column section (the remainder being wasted to control M). This section humidifies the gas stream and reduces the gas concentration to noninhibitory levels, the large liquid volume providing a stable environment for the microorganisms. Consequently, even if a section of the trickle bed is not properly wet, it will not dry out, and the microorganisms on it will not be killed.

The volume of liquid in wastewater treatment bioreactors was fixed by considerations of loading and oxygen transfer. There is no corresponding way of specifying the liquid volume in a gas-treatment bioreactor. Best practice must be to use as little as possible to keep the microorganisms wet, wash out potentially inhibitory metabolic products, and provide some buffering against inhibitory substrates. The design consequences of the above analysis are followed in the many gas-treatment bioreactors in successful operation.[23] They consist of beds of soil, compost, or other solids through which the contaminated gas is blown in a plug-flow mode. No attempt is made to remove excess biomass (i.e., $M = \infty$), and just sufficient nutrient-laden water is added to prevent the media from drying out.

5. CONCLUSIONS

The essence of engineering is design, and design for biological waste treatment is a three step process. The bioreactor configuration must first be chosen. There are a large number of possibilities, and the optimum choice depends critically on questions about the waste and the microorganisms that can be answered by laboratory experiments. What is the waste concentration ranging from pure nonaqueous liquids to a few parts per billion of contaminant, and how does it compare with the concentration (S_M) at which metabolism starts to be inhibited and the minimum concentration that can be achieved in a complete-mix bioreactor? Is the waste volatile? Is it in a liquid, in which case air stripping may be a problem in an aerobic process, or a gas stream? Is it a nutrient for the biomass, or must it be degraded cometabolically? If the latter, then what other nutrients must be provided, and how much do they cost? (This controls how efficiently they must be used and how much biomass we can afford to waste.) Does the problem include a waste constituent or metabolic product that may biosorb and/or inhibit the biomass?

Several bioreactor types have been suggested. For pure nonaqueous liquids a fed-batch type may be optimal because wasted biomass is the only effluent. For slightly contaminated groundwaters, it may be sufficient to encourage the growth of appropriate microorganisms in sand filters. A complete-mix reactor followed by a trickling filter may be optimum for wastewaters containing volatile and inhibitory constituents. For anaerobic processes, upflow sludge blanket reactors approach the theoretical upper limits of productivity due to their very high cell concentrations. For gas streams, the soil bioreactor works well for low concentrations of hydrocarbons, etc., but more advanced, trickle bed designs may be needed for high concentrations or for wastes that generate potentially inhibitory products (e.g., Cl^-).

The second step is to identify and specify approximate values of the main process variables. This is done by mathematical modeling. Virtually all bioreactors contain a gas phase and an aqueous phase that contains the microorganisms. Nutrients enter and leave the reactor in both phases, and the contaminant may be in either (or both). In the analysis given here, concepts of loading and inlet nutrient concentration are generalized to include both gas-treatment and wastewater-treatment bioreactors. The limiting nutrient is identified as the one with the lowest value of L/q, where L is the generalized loading and the specific uptake rates for the various nutrients (q) are related by the yield equations. As a general principle, either the contaminant or the most expensive added nutrient must be the limiting nutrient in each phase. The size of the reactor is then actually fixed by the gas/liquid mass-transfer characteristics of the system. For example, in conventional wastewater treatment, the contaminant is limiting in the aqueous phase (or other nutrients such as NH_4^+ and PO_4^- must be added until it is), and oxygen is the limiting (only) nutrient in the gas phase. The contaminant loading and thus the volume of the bioreactor needed to prevent anaerobic conditions are controlled by the value of k_La for oxygen. The same approach applies to cometabolic situations, although it is more difficult to apply since the yield equations have one more degree of freedom.

The single most important feature of the kinetics of bioprocesses is that they are autocatalytic. Microorganisms not only catalyze the desired reactions, but reproduce themselves as part of the process. By retaining all cells in the reactor, either by cell recycle or cell immobilization, the biomass concentration and thus the volumetric reaction rate can be increased to almost any desired level. The only limits are provided by the rates of gas/liquid transfer of nutrients or (for anaerobic processes) the appearance of imperfect mixing and mass-transfer limitations in very dense suspensions of biomass. The importance of retaining biomass in the reactor makes the mean cell residence time, M, the single most important process variable. For a complete-mix system, the value of M, together with the inherent microbial kinetics, fixes the steady-state contaminant concentration in the liquid phase and the balance between nutrient consumption for growth and maintenance. A high value of M means a high cell concentration in the bioreactor, a small amount of wasted biomass for disposal, and a low effluent contaminant concentration, conditions that become increasingly vital as the waste becomes more dilute, and thus the amount of biomass that can be grown on it decreases.

The third step is the construction and operation of a pilot plant in order to finalize the design and fix values of the variables for the full-scale process. This step is vital for each new waste to be treated, but it must not be forgotten that a pilot plant can study only a single bioreactor design and a relatively narrow range of process variables. When developing a new bioprocess technology, it is highly improbable that a pilot plant will produce the optimum solution to the design problem unless Steps 1 and 2 have been carefully considered first.

6. ACKNOWLEDGMENT

This work was supported under contract no. DE-AC07-76IDO1570 from the U.S. Department of Energy to the Idaho National Engineering Laboratory/EG & G Idaho, Inc., with funding from the Exploratory Research and Development Program.

7. NOMENCLATURE

a	Gas/liquid interface area per unit volume of liquid
A	Amount biosorbed per C-equiv. of cells
B	$k_L a/QH$: mass-transfer effectiveness number
C	Gas phase concentration
F	Liquid flow rate
H	Henry's law constant
K	Monod half-velocity constant
k	ATP required for maintenance per hour per C-equiv. of biomass
k_L	Gas/liquid mass-transfer coefficient
k_m	Cell maintenance coefficient
L	Generalized loading = $k_L a S^* + S_i/Y$
M	Mean cell residence time
q	Specific rate of consumption or production of a component
$\bar{q}$	Maximum specific rate of consumption or production of a component
Q	Gas flow rate (vol/vol liquid · min)
S_{in}	Effective inlet concentration = $(S_i + k_1 a \tau S^*)/(1 + k_1 a \tau)$
S	Concentration in liquid
S^*	Equilibrium concentration in liquid
t	Time
V	Liquid volume in system
X	Cell concentration in liquid (C-equiv./L)
δ	Intracellular energy parameter (see Equation 10)
γ	Available electrons per mole or C-equiv.
μ	Specific biomass growth rate
$\hat{\mu}$	Maximum specific growth rate
τ	Liquid residence time

8. SUBSCRIPTS

e	Effluent
i	Influent

o   Oxygen
A   Auxiliary nutrient
B   Biomass
M   Metabolic product
N   Nitrate
R   Respiratory product
g   Gas-phase limiting nutrient

9. REFERENCES

1. Erickson, L. E., and D. Y. Fung. *Handbook on Anaerobic Fermentations* (New York: Marcel Dekker Inc.,1988).
2. Andrews, G. F. "Estimating Cell and Product Yields," *Biotechnol. Bioeng.* 33:256–265 (1989).
3. Grady, C. P., and H. C. Lim. *Biological Wastewater Treatment* (New York: Marcel Dekker Inc.,1980).
4. Andrews, G. F. "Fluidized-bed Bioreactors," *Biotechnol. Genet. Eng. Rev.* 6:151–178 (1988).
5. Andrews, G. F. "Aerobic Wastewater Process Models," in *Biotechnology, Vol. IV*, H. J. Rehen and G. Reed, Eds. (New York: VCH, 1991), pp. 407–439.
6. Rogers, R. D., J. H. Wolfram, and D. Higdem. "Microbial Processing of Volatile Organics in Industrial Waste Streams, Toxic and Hazardous Substances," *Chemical J.* (in press).
7. Leson, G., and A. M. Winer. "Biofiltration: An Innovative Air Pollution Control Technology for VOC Emissions," *J. Air Waste Manage. Assoc.* 41:1045–1054 (1991).
8. Hill, G. A., M. E. Tomusiak, B. Quail, and K. M. Van Cleave. "Bioreactor Design Effects on Biodegradation Capabilities of VOCs in Wastewater," *Environ. Prog.* 10:147–153 (1991).
9. Strandberg, G. W., T. L. Donaldson, and L. Farr. "Degradation of TCE and Trans-1-2-DCE by a Methanotrophic Consortium in a Fixed-Film Packed-Bed Reactor," *Environ. Sci. Technol.* 23:1422–1425 (1989).
10. Andrews, G. F. "The Large-Scale Bioprocessing of Solids," *Biotech. Prog.* 6:225–230 (1990).
11. Ferguson, C. R., M. R. Peterson, and T. H. Jeffers. "Removal of Metal Contaminants from Wastewaters Using Biomass Immobilized in Polysulfone Beads," in *Biotechnology in Materials and Metals Processing*, B. Scheiner, F. Doyle, and S. K. Kawatra, Eds. (Littleton, CO: Society of Mining Engineers, 1989), pp. 193–199.
12. Walker, J. F., M. V. Helfrich, and T. L. Donaldson. "Biodenitrification of Uranium Refinery Wastewaters," *Environ. Prog.* 8:97–101 (1989).
13. Tarutis, W. J., and R. F. Unz. "Behaviour of Iron and Manganese in the Sediment of a Wetland Subjected to Acidic Mine Drainage," in *Biohydrometallurgy: Proceedings of the International Symposium, Jackson Hole, Wyoming, August 13–18, 1989*, J. Salley, R. G. L. McCready, and P. L. Wichlacz, Eds. (1989), pp. 349–362.
14. Bouwer, E. J., and P. L. McCarty. "Transformations of Halogenated Organic Compounds under Dentrification Conditions," *Appl. Environ. Microbiol.* 45:1293–1296 (1983).

15. Myler, C. A., and W. Sisk. "Bioremediation of Explosives-Contaminated So. *Symposium on Environmental Biotechnology*, University of Tennessee, Knoxville (1990).

16. Oldenhuis, R., R. L. Vink, D. B. Jansen, and B. Witholt. "Degradation of Chlorinated Aliphatic Hydrocarbons by *Methylosinius trichosporium* OB-3b," *Appl. Environ. Microbiol.* 55:2819–2826 (1989).

17. Henze, M., C. P. Grand, W. Gesyer, G. V. Marais, and T. Matsuo. "A General Model for Single Sludge Wastewater Treatment Systems," *Water Res.* 21:505–515 (1987).

18. Atkinson, B., and F. Mavituna. *Biochemical Engineering and Biotechnology Handbook* (New York: Nature Press, 1983).

19. Forrest, W. W., and D. J. Walker. "The Generation and Utilization of Energy During Growth," *Adv. Microbial Physiol.* 5:213–240 (1971).

20. Stouthammer, A. H., and H. W. Verseveld. "Microbial Energetics Should be Considered in Manipulating Metabolism for Biotechnological Purposes," *Trends in Biotechnol.* 5:149–155 (1987).

21. Lawrence, A. W., and P. L. McCarty. "Unified Basis for Biological Treatment and Design," *J. Sanit. Eng. Div. ASCE* 96:757–778 (1980).

22. Hickey, R. F., D. Wagner, and G. Mazewski. "Combined Biological Fluid Bed-Carbon Adsorption System for BTEX Contaminated Groundwater Remediation," in *Fourth National Well Water Assoc. Action Conference on Aquifer Restoration, Groundwater Monitoring, and Geophysical Methods*, Las Vegas, NV (1990).

23. Fan, L. S., K. Kigie, T. R. Long, and W. T. Tang. "Characteristics of a Draft Tube Gas-Liquid-Solid Fluidized Bed Bioreactor for Phenol Degradation," *Biotechnol. Bioeng.* 30:498–504 (1987).

24. Baliga, K. Y., and J. T. O'Connor. "Biologically Mediated Changes in the Filtration of Aerated Groundwater," *J. Am. Water Works Assoc.* 63:292–297 (1971).

25. Apel, W. A., P. R. Dugan, M. R. Wiebe, E. G. Johnson, J. H. Wolfram, and R. D. Rogers. "Bioprocessing of Environmentally Significant Gases and Vapors: Methane, Trichloroethylene and Xylene Using Gas Phase Bioreactors," in *ACS Symposium Series #518, Emerging Technologies in Hazardous Waste Management*, D. W. Tedder and F. G. Pohland, Eds. (American Chemical Society, 1993).

CHAPTER 8

Regulatory Considerations

Sue Markland Day*

1. INTRODUCTION

Previous chapters have described a myriad of potential and proven environmental biotechnology applications for waste remediation and pollution prevention. The purpose of this chapter is to place the "science" of biotechnology into today's regulatory structure, to define the constraints imposed on this waste management methodology by governments, and to explore the markets for biology-based waste treatment created by sociopolitical mandates. The chapter will begin with an overview of federal regulations that apply to bioremediation, then proceed with a more detailed analysis of the key rules as they impact specific biodegradation methodologies, and close with the observation that the decision of which regulations apply to any single bioremediation site is ultimately that of the individual regulator who is in charge of a particular site.

* Present address: Acting Associate Director, University of California Systemwide Biotechnology Research and Education Institute, San Francisco, CA.

0-87371-613-2/94/$0.00+$.50
© 1994 by Lewis Publishers

2. AN OVERVIEW OF THE FEDERAL PROGRAM

2.1 Statutes

Currently, there are four key federal laws that impact biology-based waste treatment: the Resource Conservation and Recovery Act (RCRA); the Comprehensive Environmental Response, Compensation, and Liability Act (CERCLA); the Clean Water Act (CWA); and the Toxic Substances Control Act (TSCA). Three other laws, the Clean Air Act (CAA), the National Environmental Protection Act (NEPA), and the Federal Insecticide, Fungicide, and Rodenticide Act (FIFRA), also play a role in the use of biotreatment. As new regulations are promulgated for the 1990 amendments to the CAA, this environmental statute will grow in its importance by significantly expanding environmental biotechnology applications to gas-phase pollutants and should assume a position in the "key" federal regulation category within the next few years.

A brief summary of each act can be found in Section 7. Each of these laws, based on its congressional mandate, addresses either the application of biology-based waste treatment or the microorganism(s) utilized in the waste management process.

Federally mandated control of pollution and clean-up processes has created markets for environmental remediation technologies. In most cases, both direct and indirect demands for biotreatment methodologies have been generated because of two recent statutory language trends. The first appearance in federal laws of statutory language was in 1984. These laws emphasized permanent treatment technologies PL 98-616, the 1984 Hazardous and Solid Waste Amendments (HSWA) to RCRA, which states:

> The Congress hereby declares it to be the national policy of the United States, that, wherever feasible, the generation of hazardous waste is to be reduced or eliminated as expeditiously as possible. Waste that is nevertheless generated should be treated, stored or disposed of so as to minimize the present and *future* threat to human health and the environment.

The permanency trend continued with the 1986 amendments to CERCLA (PL 99-499). The 1986 Superfund Amendments and Reauthorization Act (SARA) states "[t]he President shall select a remedial action that is protective of human health and the environment, that is cost effective, and *that utilizes permanent solutions* and alternative treatment technologies ... to the maximum extent practicable." Section 121 of SARA directs that "[r]emedial actions in which treatment which permanently and significantly reduces the volume, toxicity, pollutants, and contaminants is a principal element, are to be preferred over remedial actions not involving such treatment." The Federal Water Pollution Control Act, as amended in 1987 (PL 100-4), declares "it is the national policy that the discharge of toxic pollutants in toxic amounts be prohibited."

The Pollution Prevention Act of 1990 makes pollution prevention and reduction a national policy: "...pollution that cannot be prevented or recycled should be treated in an environmentally safe manner whenever feasible; and disposal or other release into the environment should be employed only as a last resort." A primary goal of the 1990 CAA is "to encourage or otherwise promote reasonable federal, state, and local governmental actions, consistent with the provisions of this Act, for pollution prevention" (Section 101(c) of PL 101-549).

In addition to making permanent treatment a national priority, recent environmental legislation directs the source of the pollution to reduce their waste generation. Statutory language in amendments to the CWA and Solid Waste Act directs the certification by the waste generator of company efforts to decrease waste production at its source. For example, the CWA states in its preamble that "...the industrial user shall certify that it has a program in place to reduce the volume or toxicity of hazardous wastes generated to the degree it has determined to be economically practical." The HSWA language also required generators to attempt waste reduction. Under those amendments, hazardous waste generators are required to certify on their hazardous waste manifests and annual permit reports that they have a program in place to reduce the volume or quantity and toxicity of their hazardous wastes as much as economically practical. The RCRA rules also require generators to describe on their biennial reports the efforts they have undertaken during the year to reduce the volume and toxicity of their hazardous waste and to compare these efforts to previous years. The Pollution Prevention Act of 1990 adds source reduction, recycling, and waste treatment data to the Title III reporting requirements mandated by SARA.

The congressional mandate is clear. Industrial processes that generate less waste and waste management methodologies that detoxify or breakdown to primary elements pollutants are the preferred choices. One biotreatment industry representative identified this waste reduction mandate as an unexpected bonus in the environmental amendments signed into law in the 1980s.[1] As other authors of this book have demonstrated, biology-based waste treatment can fulfill the objectives of waste detoxification and reduction for many soil, water, and air contaminants. Permanency is the promise of bioremediation. Microorganisms can convert the select toxic substances into carbon dioxide, water, and cell mass: a truly permanent waste treatment. For completely biodegradable toxic substances, no poisonous residuals remain to manage. If not total destruction, then environmental biotechnology may be able to detoxify or sequester pollutants. Biotransformation can convert hazardous chemicals into more benign, less toxic compounds for wastes that are not completely mineralized by biological actions. For metals and radioactive wastes, which, unlike organics, cannot be destroyed by incineration, microorganisms may be used as a biological sponge to capture these toxic compounds for reuse or to concentrate wastes into smaller volumes for disposal.

2.2 Regulations

Environmental laws are written to provide congressional direction to regulatory agencies. The mechanism by which the congressional goals and objectives are communicated to the public is in the form of regulations issued primarily by the U.S. Environmental Protection Agency (EPA). In addition to promulgated regulations, the U.S. EPA also produces guidances (documents that describe what the regulations mean and how to implement them) and individual letters or memos addressing specific questions from the public, the regulated universe, state regulatory agencies, or EPA regional offices. These documents, plus administrative law court decisions, provide the framework for the application of biotechnology to waste management.

As mentioned earlier, biology-based waste treatment is controlled under federal regulations based on the microorganisms themselves and the application of the microbes to contaminated soils, water, or air. The latter regulatory structure applies to all treatment methodologies including nonbiological treatment like incineration, while the former only impacts treatment processes that utilize microorganisms.

2.2.1 Regulations Affecting the Microorganism

Since the majority of applications of biotechnology to hazardous waste require the organism or products derived from or made by the microorganism, the regulations addressed initially in this chapter are those which control the review process for individual microorganisms and bioproducts such as nonliving biomass or enzymes produced by the organisms. Three rules fall into this category: the Toxic Substances Control Act (TSCA), the Plant Pest Control Act, and the Federal Insecticide, Fungicide and Rodenticide Act (FIFRA). Any bacteria, fungi, virus, or product consisting of attenuated microorganisms utilized in the environment can be regulated under TSCA. The other two regulations address smaller universes of microorganisms impacting relatively few of the classic bioremediation microbial workhorses.

The TSCA enables the EPA to gather information on chemical risks from those who manufacture and process chemicals. For the purpose of TSCA regulation, the EPA has defined microorganisms (both dead and alive) and compounds produced biologically as chemicals. Under this statute, the EPA may regulate a chemical's unreasonable risks at any stage in its lifecycle: the manufacturing, processing, distribution in commerce, use, or disposal. The TSCA is a notification statute. A 90-day advance contact with the agency, called a pre-manufacturing notice (PMN), is required of all the manufacturers or importers of new chemicals. Any chemical is considered "new" if it is not listed on the TSCA Inventory of existing chemicals or expressly excluded from coverage by the statute. Even though a chemical is already listed on the Inventory, additional uses may be considered by the EPA to be significant and require a 90-day notification (significant new use notice or SNUN).

The TSCA process works as follows. At least, 90 days before a firm initiates the importation or production of microorganisms or compounds derived through microbial action, a PMN must be submitted to the EPA's Office of Toxic Substances for review. Upon receipt and certification that the application is complete, federal regulators review submitted data including health and safety information provided by the firm and make a determination whether the microorganism or product presents any unreasonable risks.[2] The microorganism or bioproduct is then listed on the TSCA Inventory, and commercialization is allowed. For some compounds, the TSCA regulators may determine that selected uses of the product or organism is acceptable, but for other uses, a new approval process, called the SNUN, is required. Under the existing TSCA regulations, small quantities of product and university, federal, and private sector research are exempted. In practice, naturally occurring microorganisms are exempt from reporting under TSCA.

In 1986 with the publication of the *"Coordinated Framework for Regulation of Biotechnology"*, the U.S. EPA alerted the public that biotechnology products and processes (that were not covered by USDA or FDA and were not regulated by the EPA under the FIFRA) would be included under the TSCA.[3] At that time, the Office of Toxic Substances, responding to what they considered political pressure, initiated a rule-making process specifically for microorganisms.[4] This TSCA biotechnology rule is still under development with a proposed rule projected to be published in the summer of 1992.

As of February, 1992, no genetically engineered microorganism for hazardous waste treatment had been submitted for approval under TSCA. This statement does not mean that unicellular creatures are not currently being employed for waste remediation. It only means that the TSCA microbial agent rule promulgation has moved away from total coverage and appears to be focusing on a narrow universe consisting of only genetically modified microorganisms introduced intentionally into the environment.

In the summer of 1991, the Office of Toxic Substances published a *"Federal Register"* notice offering its latest TSCA draft rule for public review. Since its first draft rule circulated in the spring of 1989, the coverage of the TSCA requirements had narrowed. No longer did the U.S. EPA propose to require notification when a naturally occurring microorganism was utilized or sold for environmental clean-up. In fact, the most recent draft grandfathers all bacteria, et al. into the system that have not been genetically modified by man's direct intervention. This draft has abandoned earlier regulatory attempts to capture pathogenic, naturally occurring organisms for EPA review. The EPA now proposes to include implicitly all naturally occurring microbes on the TSCA Inventory.

In contrast to TSCA requirements for chemicals, the 1991 draft biotechnology rule expands the notifiers to include small businesses, universities, and federal research laboratories. The EPA has chosen to focus exclusively on microorganisms created through genetic manipulation freeing those individuals relying on naturally occurring organisms (both indigenous and nonindigenous) from TSCA regulation.

Although today's field applications of naturally occurring microorganisms may be excluded under TSCA, a basic understanding of the proposed regulation is useful. The draft *"Microbial Products of Biotechnology: Proposed Regulation under the Toxic Substances Control Act"* requires notification of the EPA when microorganisms that have been modified by alteration of their genome are deliberately released into the environment. As discussed earlier, naturally occurring microorganisms are considered to be implicitly included on the TSCA Inventory, therefore, not regulated under the draft TSCA biotechnology rule. In contrast to the treatment of chemical products under TSCA, small quantity and university activities that utilize bioengineered microorganisms will not be exempted from the TSCA reporting requirements. A Microbial Commercial Activity Notice (MCAN) is introduced to replace the PMN. A TSCA Environmental Release Application (TERA), described as a fast-track MCAN, is introduced for research activities. The draft rule includes exemptions for specific genetically manipulated organisms, but those listed are not commonly used for environmental restoration.

The draft TSCA rule identifies properties of the introduced genetic material that the EPA believes would decrease the environmental impact of an introduced microorganism. These characteristics are described below.

- The genetic material should be limited in size and must consist only of the following: (1) the structural gene(s) of interest; (2) the regulatory sequences permitting the expression of solely the gene(s) of interest; (3) associated nucleotide sequences needed to move genetic material including linkers, homopolymers, adaptors, transposons, insertion sequences, and restriction enzyme sites; (4) the nucleotide sequences needed for vector transfer; and (5) the nucleotide sequences needed for vector maintenance.
- The genetic material must be well characterized. The function of the products expressed from the structural gene(s), the function of sequences that participate in the regulation of expression of the structural gene(s), and the presence or absence of associated nucleotide sequences should be known.
- The genetic material must be poorly mobilizable. For conjugation, the transfer frequency must be less than 10^{-8} transfer events per recipient microorganism; for transduction, frequency must be less than 10^{-8} transfer event per bacteriophage; and for natural transformation, less than 10^{-8} transfer events per recipient microorganism.
- The genetic material must be free of certain sequences (nucleotide sequences that encode for certain toxins are listed).

In addition, the draft identifies the information considered necessary about the microorganism, the site where it is to be applied, and the target compounds that it was designed to degrade (Table 1). Much of this data is critical for the optimization of a bioremediation whether the organisms are naturally occurring or genetically manipulated.

The microorganisms used for waste remediation are also subject to four other federal regulations. The Federal Insecticide, Fungicide and Rodenticide Act (FIFRA), enforced by the U.S. EPA's Pesticide Office, regulates the use of microorganisms as pesticides. In 40 CFR Subpart A 152.3 (s), a pesticide is defined as "any substance or mixture of substances intended for preventing, destroying, repelling, or mitigating any pest... ." An organism (except man) is defined as a pest under circumstances that make it deleterious to man or the

Table 1.   Information Required by Draft TSCA Biotechnology Rule

1. Identity of the organism
2. List of published health and environmental data on microbe plus its unmodified parent
3. Summary of unpublished health and environmental data
4. An estimate of the amount of the organism to be produced annually (including expected viability)
5. Information on the sites where the organism is manufactured, processed, used, and disposed of
6. Worker exposure information
7. Estimated amount of the organism to be released annually
8. Detailed description of transport parameters
9. Description of intended use including all engineering controls and personal protective equipment
10. Information on locations, conditions, and purpose of intended release; anticipated survival, multiplication, and dissemination; identification of target organisms, if any; the anticipated biological interactions with and effects on target organisms as well as other organisms; a description of host range; involvement in biogeochemical or biological recycling processes; etc.
11. Description of intended disposal (e.g., efficacy of inactivation)
12. Exposure potential and dispersal routes
13. Other information that the submitter believes necessary for risk assessment

environment. Competing with indigenous organisms could fall under this definition. For example, if a company advertises that its commercial bacteria, fungi, et al. will displace existing populations of nondegradative organisms and suggests that the latter are deleterious to successful degradation, then the commercial product might be considered a pesticide. It is the product's label and marketing materials that govern whether FIFRA will apply to microorganisms or consortia of microorganisms used to degrade hazardous wastes.

The use of any biological additive in the navigable waters of the United States including territories and adjoining shorelines, the waters of the contiguous zone, and the high seas is regulated by Subpart H of the National Contingency Plan (NCP), the regulations supporting the federal Superfund program. To have a microorganism or a mixture of microbes listed on the NCP Product Schedule, the producer has to follow steps delineated in 40 CFR 300.86. On March 8, 1990, new regulations became final for the NCP, which allow the authorization of the use of biological additives including products not listed on the NCP Product List when the use of the product is necessary to prevent or substantially reduce a hazard to human life.

The third regulation applicable to the use of microorganisms can be found in the Plant Pest Act. This rule, which is enforced by the U.S. Department of Agriculture, requires prior approval by the Department of Transportation for potential plant pathogen shipments across state lines. Since many of the soil microorganisms utilized in bioremediation have cousins that are agricultural pests, this rule might apply to soil or bacteria/fungi movements. The sender/receiver has the responsibility to demonstrate that the particular microbe is not a plant pathogen. If a member of the same genus is a plant pathogen, then the microorganism is considered a possible suspect. Relatives of soil bacteria, the workhorses of bioremediation, could fit this criteria. Periodically, microorganisms exempted from the Plant Pest Act are published in the *Federal Register*.

Another microorganism-specific regulation has been promulgated by the Department of Transportation. On January 3, 1991, a final rule was promulgated that delineates packaging and shipping requirements for etiologic agents. The regulation was created in the wake of public fear regarding exposure of postal employees to AIDs. However, if interpreted broadly, the regulation could have an unexpected impact on biology-based waste treatment. A comprehensive definition of etiologic agents captures most bacteria, viruses, protozoa, and fungi; therefore, if enforcement is based on the universal potential of microorganisms to cause disease, then the shipment of microorganisms will become difficult.[5]

A similar problem arises under the Medical Waste Tracking Act (Subpart J of HSWA) when one acknowledges the potential of microorganisms to cause infection. Under this federal pilot study (only applies to New Jersey, New York, Connecticut, Rhode Island, and Puerto Rico), any off-site shipment of discarded microorganisms would require a manifest. This requirement only goes into effect if the spent cell mass is shipped from its point of generation. Moreover, disposal of exhausted biomass on-site and off-site is further complicated under Subtitle C of the Resource Conservation and Recovery Act (RCRA), which requires a federal register delisting notice to dispose of any compound (or microbe) that has been mixed with a hazardous waste.[6]

The last regulation addressing the organism, not the process by which it is used, is included in the Food, Drug, and Cosmetic Act and enforced by the Food and Drug Administration. This rule requires that all microbial products used in food service facilities be certified pathogen free. A companion rule is administered through the USDA for food processing plants. Although the applications for biodegradation in this book have focused on the treatment of hazardous compounds, one of the classic uses for biotreatment has been and continues to be the use of microorganisms to degrease sewer lines in food processing and retail establishments.

Of significant note, the seven regulations described above do not emphasize efficacy demonstrations. Each one addresses safety issues as they relate to public health and the environment, but do not ask whether specific microorganisms can degrade target compounds.

2.2.2 Regulations Affecting the Application of Microorganisms

Not all waste stream management has been subject to a federal regulatory process. Historically, this unregulated universe has been the growth industry for biology-based waste treatment. For example, cresotes and petroleum bottom sludges were not considered hazardous wastes until 1990, yet land farming relying on indigenous microorganisms had been used to reduce waste volume for these compounds for years. In the past, many of the compounds identified for pretreatment under the recent CWA amendments were unregulated if a firm's sewer outfall connected directly to the local municipal wastewater treatment plant. Yet, food processing plants and animal feed lots have long standing treatment ponds teeming with microbial life, and the wastes generated by these commercial activities have historically been recognized as biodegradable. High sulfur coal has been routinely used as an energy source. Only under the most recent CAA amendments has sulfur dioxide been identified as a priority pollutant. Prior to the 1990 CAA amendments when constructing their facilities, coal gasification plants routinely installed cooling towers. Sulfur-reducing microbial consortia, which colonize the cooling towers, were recognized as pollutant utilizers.

Leaking underground storage tanks were captured under federal regulation for the first time in 1984. Until that time, the clean-up of product-tainted soils around thousands of this nation's gasoline stations was not covered by federal regulation. Contamination of local environments, however, encouraged publicly minded gas station owners to explore bioremediation before passage of HSWA. Although acid mine drainage appears to be captured by the CWA, it is still unclear if government regulates mining wastes under the solid waste laws. Mining company operators, aware that indigenous populations of microorganisms have bloomed in the runoff from these mines, have recognized the role of microorganisms in decreasing the destruction caused by acidic effluents. In short, the applications for bioremediation have a long history. On the other hand, the constraints now placed on biological treatment by prescriptive federal environmental regulations are relatively new.

For many waste streams, the biological treatment process, which is similar to physical and chemical methodologies, is controlled by permits or compliance orders issued by the federal U.S. EPA or by a state program authorized to act as an agent of the federal government. These procedures apply to all treatment regimes and will be explained briefly below. An analysis of these process controls as they impact the application of bioremediation is presented after the regulatory overview.

2.2.3 Resource Conservation and Recovery Act

The Resource Conservation and Recovery Act (RCRA) is the primary federal regulation that applies to the treatment of hazardous wastes. The RCRA regulations control hazardous wastes under Subtitle C, nonhazardous wastes in Subtitle D, underground storage tanks in Subtitle I, and medical wastes in Subtitle J.[7] Hazardous waste is defined in the RCRA as a solid waste or combination of solid wastes, which because of its quantity, concentration, or physical, chemical, or infectious characteristics may: (1) cause, or significantly contribute to, an increase in mortality or an increase in serious irreversible or incapacitating reversible illness; or (2) pose a substantial present or potential hazard to human health or the environment when improperly treated, stored, transported, or disposed of, or otherwise managed. All other solid wastes, except medical wastes, are considered to be nonhazardous and are most commonly disposed of in a sanitary landfill. Although most subtitles of RCRA affect bioremediation applications, the focus of this chapter will be limited to the hazardous waste regulatory requirements.

The RCRA, when signed into law in 1976, set three major goals: (1) to protect human health and the environment, (2) to reduce waste and conserve energy and natural resources, and (3) to reduce or eliminate the generation of hazardous waste as expeditiously as possible. It is important to note that RCRA establishes a framework for the proper management of solid waste, both hazardous and nonhazardous, but does not address the problems of hazardous waste encountered at inactive or abandoned sites. In general, RCRA covers both "new" wastes and the clean-up of releases from "old" wastes generated at hazardous waste treatment, storage, and disposal facilities, while the Comprehensive Environmental Response, Compensation, and Liability Act (CERCLA) only addresses "old" wastes found at abandoned sites.

Subtitle C of the RCRA establishes a cradle to grave hazardous waste management program with regulations regarding the generation, transportation, and treatment, storage, or disposal of hazardous wastes. The regulated universe is defined by two mechanisms: the waste's *characteristic* and the *listing* by the agency of specific or groups of wastes.[8] If a hazardous waste is intermingled with nonhazardous wastes including microbial cell mass, then the entire mixture is considered a hazardous waste.

For those wastes captured under Subtitle C, the U.S. EPA sets administrative requirements for three categories of hazardous waste handlers: generators, transporters, and owners or operators of treatment, storage, and disposal facilities (TSDs). Technical standards for the design and safe operation of TSDs and acceptable contaminant levels after treatment and appropriate treatment technologies are also included in the RCRA rules. The TSD regulations serve as the basis for developing and issuing the permits that each hazardous waste facility is required to obtain.

2.2.4 Regulations Affecting the Treatment Facility

Biotreatment of RCRA regulated wastes at a commercial off-site treatment facility requires a RCRA permit. A location specific RCRA permit is required for on-site treatment. Such a permit is necessary when the treatment facility is located at a non-CERCLA site utilizing either a permanent facility or a mobile treatment unit. Hazardous waste treatment pilot studies require a RCRA research, development, and demonstration (RD&D) permit. The RD&D procedure is outlined in a guidance issued in the early 1980s for hazardous waste treatment. A Treatability Exemption Rule has been finalized that allows the treatment of small quantities of hazardous waste without a permit. This rule allows the development of bioremediation procedures at bench scale without the delay experienced when obtaining a RCRA permit.[9] *In situ* treatment does not require a RCRA permit, but the treatment procedure may be cited in a RCRA corrective action order. Such orders may contain the same level of operational or performance detail commonly found in hazardous waste permits.

RCRA Subtitle C permits are required for all treatment, storage, and disposal facilities managing hazardous wastes. Under RCRA, permit requirements are specified for different types of treatment units. For contained treatment systems (such as large bioreactors), the requirements are most promulgated as tank standards. Tanks, under the RCRA rules, are stationary devices designed to contain an accumulation of hazardous waste and constructed primarily of nonearthen materials. There are distinct advantages (regulatory avoidance) when the biological treatment unit is a tank. As described later in the chapter, the constraints on bioremediation imposed by the RCRA Land Disposal Restrictions (LDRs) do not apply to tank-based treatments.

In addition to tanks, biological treatment may be done in four other land-based configurations that have specific RCRA permit requirements. The use of the following structures for biotreatment must comply with the RCRA LDRs. A surface impoundment is a depression or diked area (e.g., pit, pond, or lagoon) used for storage, treatment, or disposal. The impoundment is open on the surface and designed to hold an accumulation of waste in liquid or semi-solid form. Historically, biotreatment of petroleum primary refinery wastes in lagoons has been carried out in surface impoundments. Waste piles, used for treatment or storage of a noncontainerized accumulation of solid, nonflowing hazardous waste, are regulated under 40 CFR Part 264 Subpart L. Waste piles used for disposal must comply with Subpart N requirements. Composting utilizes waste piles.

Subpart M impacts land treatment processes, while 40 CFR Part 264 Subpart N contains the requirements for landfills. Land treatment is the process of using soils and microorganisms as a medium to biologically treat hazardous waste. As with surface impoundments, land treatment units and hazardous waste landfills must install double liners and leachate collection systems.[10]

The following sections describe key elements of the federal hazardous waste program enforced through the RCRA Subtitle C permit system.

2.2.5 Corrective Actions

Unique from current federal water and air regulations is the Corrective Action (CA) provision introduced into the 1984 amendments to RCRA, the Hazardous and Solid Waste Amendments (HSWA). The proposed CA rule was published on July 27, 1990. In order to maintain a hazardous waste permit, the treatment, storage, and/or disposal (TSD) permit holder must include corrective action measures, which commit the firm to the remediation of contaminated solid waste management units on the site. Consequently, the HSWA corrective action provisions have created a clean-up program parallel to the federal Superfund managed under the RCRA hazardous waste program. The management of contaminated soils and groundwater restoration, until HSWA, was the responsibility of the EPA Superfund managers. Now the CA provisions bring releases from selected "old" waste under the RCRA managers.

This dual program is significant to the application of biology-based waste treatment methodologies. If contaminated soil and debris are located at a facility that holds a RCRA Subtitle C permit, then the remediation strategy must comply with the Land Disposal Restrictions (LDRs). If the pollution is located at a federal

Superfund site, then the clean-up is exempted from RCRA permit requirements and only has to comply with the LDRs in spirit.[11]

Another program element the HSWA LDRs has the potential to be a major constraint restricting environmental biotechnology applications to hazardous waste management. Statutory LDRs are triggered by the placement of a restricted waste in a land-based waste management unit. Minimum technology standards are triggered by the creation of new as well as the replacement or lateral expansions of existing surface impoundments or landfills. Therefore, the movement of soils and debris contaminated with hazardous waste to a treatment cell or a compost pile, which is a routine step in biology-based waste treatment, would trigger land disposal restrictions. The LDRs, when fully implemented, control which treatment process can be used and/or the clean-up level determined by the final concentration of residual waste.

2.2.6 Regulations Affecting Contaminated Soils and Groundwater

The regulations for the Comprehensive Environmental Response, Compensation, and Liability Act (CERCLA) — more commonly known as Superfund — are organized in the National Contingency Plan (NCP). This document, which was amended by new rules promulgated in February, 1990, describes the procedure for the federal site restoration program. Other clean-up programs such as those managed by the states, Indian tribes, and the federal agencies fall outside of the NCP and are not necessarily exempted from the RCRA program requirements including the LDRs.

For those sites addressed under CERCLA, innovative treatment methodologies are preferred. In fact, a program titled the Superfund Innovative Treatment and Evaluation Program (SITE) was created by the act and is considered by federal regulators to be the logical window for federal evaluation of biology-based hazardous waste treatment procedures. Permits are not required for CERCLA actions, instead Records of Decisions (RODs) are utilized and compliance orders issued. The process of remediation is negotiated between the responsible parties, the federal government, and often the state, as are the target contaminant levels for restored soils and groundwater. According to the 1986 Superfund Amendments and Reauthorization Act (SARA), the RODs and their companion compliance orders must reflect applicable, relevant, and appropriate requirements (ARARs).[12]

The National Environmental Policy Act (NEPA) is the last major environmental law that harbors requirements applying to all treatment technologies. This law requires any federally funded activity to be analyzed for its environmental impact. For example, if bioremediation is planned for a site on federal property, at least a short form Environmental Assessment document must be submitted to the U.S. EPA. To date, few federal clean-ups have relied on bioremediation, but considerations unique to biology-based treatment such as potential by-products, dispersion of nonindigenous organisms, or disruption of carbon or nitrogen cycles deserve attention in these analyses.

3. IMPACT OF REGULATIONS ON FIELD USE OF BIODEGRADATION METHODOLOGIES

Building on this overview of regulations, let us now examine, in more detail, the key regulatory compliance issues that control the field use of the biodegradation methodologies described in this text.

3.1 Type of Waste

Whether the chemical contaminant is a radioisotope, mining waste, a mixture of PCBs, priority pollutant, or listed hazardous waste may determine which regulations apply to biology-based waste treatment. Some selected compounds are addressed by name or type in a regulation's enabling legislation, while other pollutants are covered by a particular rule based on what media the waste is contaminating. Examples of the former include regulation of PCB clean-up under the Toxic Substances Control Act (TSCA), of selected mining wastes under Subtitle D of the Resource Conservation and Recovery Act (RCRA), of radionucleotides under the Nuclear Regulatory Commission, and of dioxin emissions under the Clean Air Act (CAA). An example of the latter is the coverage of benzene under both the Clean Water Act (CWA) (as a priority pollutant) and Subtitle C of the RCRA (as a listed waste) depending on whether the chemical is contaminating water or soil. In addition, sulfur is not considered a pollutant if it is found in soils, but as a sulfur oxide, the compound is listed as a priority pollutant under the CAA.

The successful application of biology-based waste treatment can be seriously impacted by which environmental regulations cover the waste of interest. Inclusion of a particular waste stream under Subtitle C of RCRA may limit the configuration of treatment units that can be utilized or may specify a clean-up level not achievable through biodegradation. For example, the passive biological treatment of primary refinery waste sludges in waste ponds and lagoons is no longer acceptable. In May, 1991, all of these sludges became listed hazardous wastes; therefore, their treatment now must comply with RCRA requirements, which in effect bans land-based treatment units.[13]

3.2 Nature of the Site

Whoever is the owner or operator of a site may determine which regulations must be complied with when using biological treatment. If the contaminated site is located on federal property, the National Environmental Protection Act is the umbrella law. If the property is abandoned and is on the National Priority List (NPL), the National Contingency Plan (NCP) for the federal Superfund program applies. If the property is abandoned, but it is not on the NPL, then state waste management programs may prevail. If the contamination is located at a facility holding a Subtitle C TSD permit, then remediation is covered under the RCRA Corrective Action Provisions of the act. If the pollutant is generated during a chemical manufacturing process, the intermediary or by-product may be captured under the Toxic Substances Control Act (TSCA).

Under Section 1004(27) of the RCRA, hazardous waste mixed with domestic sewage treated at a public owned treatment works (POTW) is not regulated. This provision is known as the "Domestic Sewage Exclusion". Since the large volumes of this nation's hazardous wastes can be found in untreated wastewater, this exclusion has created a large loophole in the treatment of toxic waste, although new amendments to the CWA may capture this exclusion under its pretreatment standards.

Farm land contamination by overuse of pesticides and fertilizers is another example of how the environmental regulations focus on certain generators (almost exclusively nonagricultural) and completely ignore others. This type of contamination may fall under the Safe Drinking Water Act's Recharge Zone provisions, but it is not an obvious candidate for CERCLA coverage. The acid mine drainage waters may be captured under the CWA, but which RCRA rules apply to the solid mining wastes are still under debate. Industrial wastes such as drilling fluids association with oil exploration do not yet appear to be covered under any federal regulation.

3.3 Location of the Waste

If the pollution is nonpoint source in nature, then portions of the CWA and Safe Drinking Water Act may be involved. Contaminated river sediments are controlled under the CWA. If a contaminant is located in the feedstock for an incinerator, its final destination will determine which regulations apply. Emission standards for the CAA apply to combusted materials (the gases), while the solid waste disposal rules take precedence when hazardous materials such as heavy metals remain in the ash.

Different regulations come into play for pollutants in air, water, or soil. Groundwater and surface water contamination are also segregated into distinct rules. Whether the wastewater is directly discharged from a company's facility through an outfall or whether the waste is discharged into a POTW also modifies the requirements for waste management. Air regulations are often set based on the size of the incinerator or on its age, with the smaller units exempt in some states. This lack of coverage could limit the application of biology-based pollution control equipment if biological air treatment units only work effectively on the class of equipment that is exempted due to small size.

The choice of whether to excavate a contaminated soil or to do an *in situ* remediation is impacted by regulations. The RCRA LDRs, in effect, require the construction of a permitted TSD before lifts of contaminated soils can be treated, but do not apply if one adds bacteria, fertilizers, and oxygen to untouched soils. However, to avoid coverage under the RCRA LDRs, all of these ingredients must be introduced in pristine water, not by recycling contaminated groundwater from the site.

3.4 Waste Treatment Unit

Federal regulation citations impact the type of biological treatment units utilized. Local and state zoning and building codes may also apply, but are not covered within the scope of this chapter. Waste treatment units range in design from small on-site bioreactors to large open land areas with sprinkler systems. Sewage treatment facilities, compost piles, holding ponds, cooling towers, and air filters serving as homes to biofilms have all been used in bioremediation projects. The fact that microorganisms are the active ingredient in all of these treatment units does not alter the basic performance and design standards for the unit, although the presence of microorganisms may result in additional regulations that address the safety of biomass.

The RCRA permits are not required of owners or operators of totally enclosed treatment facilities. According to 40 CFR 260.10, such a facility is one for the treatment of hazardous waste that is directly connected to an industrial production process and is constructed and operated in a manner that prevents the release of any hazardous waste or any constituent thereof into the environment during treatment, e.g., elementary neutralization units. While some bioreactors may qualify under this exemption, the vast majority are tanks, by the EPA's definition, and are subject to regulation.

Secondary containment and release detection systems for tanks are specified by 40 CFR Part 264 Subpart J of the RCRA hazardous waste management standards. Moreover, the subpart delineates special requirements for ignitable or reactive wastes and incompatible wastes if such wastes are treated in the bioreactor. Air and water regulations also apply to tanks. Any emissions or effluents must be controlled from a hazardous waste treatment facility.

Of particular interest to those utilizing biology-based waste treatment is the latest draft of the TSCA biotechnology regulations. Surprisingly, these regulations exempt microorganisms used exclusively in tanks from its notification requirements. However, under the RCRA mixture rule, the spent cell mass generated in a bioreactor may require "delisting" before disposal. If a RCRA listed hazardous waste is treated in the tank, then everything in that tank is considered a hazardous waste even if the original waste has been totally degraded. To dispose of the cell sludge, one would have to send the biomass to a hazardous waste treatment or disposal facility or go through a federal rule-making process to delist the residue.

Historically, organic mixtures amenable to biodegradation such as creosote and paper mill wastes have been placed in surface impoundments. Regulations specified by 40 CFR Part 264 Subpart K require impoundments to have double liners and a leachate collection system (minimum technological requirements). When new hazardous wastes are listed under RCRA Subtitle C, thereby expanding the application of minimum technological requirements to previously unregulated surface impoundments, the impoundment must be retrofitted with double liners within 4 years or closed. Four very limited variances to this requirement exist. In 1990, creosote was listed as a hazardous waste; therefore, biological treatment of this wood-preserving mixture must now comply with all RCRA requirements. It is unclear how the LDRs impact the movement of the wastes from an unlined to a lined impoundment, but the LDRs (which must be promulgated 6 months for newly listed chemicals) most certainly will specify a clean-up level or acceptable treatment methodologies for these compounds. In addition to meeting RCRA rules, disposal of the leachate will be controlled under the CWA if point source discharged and, potentially, air emissions would be regulated in a nonattainment area.

Waste piles, used for treatment or storage of a noncontainerized accumulation of solid, nonflowing hazardous waste, are regulated under 40 CFR Part 264 Subpart L. Compost piles, which treat hazardous waste, fall under these rules. This subpart requires waste piles to be lined, have a leachate collection and removal system, be protected by 25-year storm run-on and run-off control systems, and utilize a wind dispersal management system.

Although now being phased out because of the RCRA LDRs, Land Treatment, the process of using soils and microorganisms as a medium to biologically treat hazardous waste (land farming) has been the mainstay of bioremediation. The EPA Regional Administrator or the state government if authorized in lieu of the U.S. EPA specifies the design criteria including vertical and horizontal dimensions of the treatment zone in the RCRA permit. Under the LDRs, prior to land treatment, the waste must be treated to Best Demonstrated Available Technology (BDAT) levels. An alternative to BDATs is applying for a no migration exemption. Under this exemption, for each waste that will be land treated, the owner or operator must demonstrate, prior to application of the waste to the land, "that the hazardous constituents in the waste can be completely degraded, transformed, or immobilized in the treatment zone".

In situ bioremediation is a form of land treatment that does not involve the placement of contaminated soils in another location. In most cases, this type of treatment falls under a remedial or corrective action. For such sites, no RCRA permit would be required, but air and water permits may be in order if the soil is disturbed and volatile organic emissions are created or if uncontaminated water is sprinkled on the site and the runoff, containing hazardous wastes, is discharged at an outfall.

If the biology-based hazardous waste treatment unit does not match any of the configurations described above, then Subpart Q of 40 CFR 265 Chemical, Physical, and Biological Treatment Units applies. It is possible that a treatment train, consisting of linked chemical, biological, or physical components, might fall under this rule. In summary, with the exception of air pollution control equipment and wastewater treatment plants, all biological treatment units handling hazardous wastes require a RCRA permit unless the treatment is underway at a NPL site. The compliance orders accompanying the site's Record of Decision (ROD) are equivalent to a permit. If the unit is located at a site that already holds a RCRA permit, then an amendment to the existing permit can serve the same function. Even mobile treatment units require permits. Efforts have been underway since 1987 to establish a permit by rule for such units, but the push has lost momentum in the last few years.

Research and development activities involving the application of environmental biotechnology to hazardous waste remediation require RCRA RD & D permits. The 1984 Hazardous and Solid Waste Amendments (HSWA) to RCRA established a new permitting authority for the EPA: the RD & D permits for innovative and experimental hazardous waste treatment technologies or processes. The RD & D permit was designed to allow applicants to conduct experimental testing of new hazardous waste treatment technologies or processes by modifying or waiving RCRA permit applications and procedural requirements.

A regulation titled Permitting Experimental Facilities Conducting Hazardous Waste Research is still listed as an EPA ongoing regulatory development activity, but the momentum for this rule is currently at a standstill. Demand for the Experimental Facilities permitting rule was muted with the promulgation of Section 261.4 (e) Treatability Study Samples. The treatability exclusion rule exempts from a RCRA Subtitle C permit solid waste that is: (1) no more than 1000 kg of any nonacute hazardous waste; (2) 1 kg of acute hazardous waste; or (3) 250 kg of soils, water, or debris contaminated with acute hazardous waste. Under the treatability exclusion rule, certain shipping and packaging requirements must be met, and the unused portion of the waste must be returned to its point of origin.

3.5 Location of the Treatment Activity

The location of the treatment unit also provides guidance as to which regulations apply. If the unit is to be used on-site at a Superfund site, then TSD permits are not required, but the remedial action must still address the RCRA applicable, relevant, and appropriate requirements (ARARs). All other on-site structures including mobile treatment units, which handle hazardous waste, must be permitted under RCRA, the CAA, and/or the CWA. If the treatment is done *in situ*, with no structures, then a federal permit is probably not required, but it would be prudent to observe public notification and submission of data required under Subtitle C of RCRA. Off-site biological treatment units must be permitted under these acts also; moreover, transportation and manifesting requirements must be met. State and local permits may also apply to both on-site and off-site treatment units.

Biological treatment at federal facilities are not excluded from the RCRA water or air requirements. For example, in the 1990 CAA, congress specifically directs all federal facilities to comply with all federal, state, interstate, and local requirements, both substantive and procedural.[14] In addition to these major environmental statutes, federal agencies must also comply with National Environmental Protection Act (NEPA) requirements.

3.6 How Clean Is Clean?

The acceptable level for treatment residuals may be the single most important "stopper" for bioremediation. Each law establishes different criteria for its treatment end point. Some regulations promulgated under the same act have conflicting residual concentration standards for the same pollutant depending on whether the initial compliance point occurs at the point of generation (new wastes) or in abandoned soil or debris (old wastes). Each act has a clean-up goal, and many of the regulations promulgated for the environmental laws have terms for these standards. Table 2 provides a representative set of definitions.

As discussed previously, different environmental laws may apply to the same bioremediation activity. Surprisingly, hazardous waste clean-up standards may even be different under one act. For example, in a *"Federal Register"* notice published May 30, 1991, the Office of Solid Waste has indicated its interest in establishing target clean-up levels for contaminated soil and debris that differ from those presently in effect under the LDRs. Proposed rules on these residual concentration standards for debris are projected for publication in the spring of 1992 for contaminated debris and May, 1992, for soils. Until these new rules are finalized, a waste treatment methodology might have to meet two different clean-up levels for the same pollutant. The LDRs' BDAT performance-based standards apply to new wastes, while health-based target clean-up levels for soils and debris, which had been contaminated in the past (old wastes), are found in a draft EPA CA guidance. Compare the difference between the technology-based LDRs and the risk-based CA levels for soil and debris contaminated with phenol: 2.7 mg/kg and 3000 mg/kg, respectively. Although phenols are readily degradable by microorganisms, different types of soils might limit the contaminant's bioavailability making it difficult biodegradation to obtain the 1000 times more stringent LDR residual concentration set by incineration.

Important to biodegradation is the pending publication of draft quality standards, an EPA guidance document, and technical guidelines for sediments. The U.S. EPA sediment strategy will address trigger concentrations for nonionic organics first including phenanthrene, naphthalene, dieldrin, and aldrin. Criteria for metals in sediments should follow in the near future.

As to their impact on biology-based hazardous waste treatment systems, the importance of performance versus design standards cannot be over emphasized. If the pollutant residues must meet residual standards of X, Y, and Z for water, air, and soil, then biodegradation may be optimized to satisfy the clean-up levels. If the regulation specifies the use of an incinerator (as is the case with the PCB regulations), environmental biotechnology, even if proven efficient, is not an acceptable waste management approach. Performance-based standards may also negatively impact bioremediation if the clean-up levels are established by existing technology such as chemical fixation and/or incineration.

As described throughout this chapter, the biological treatment of hazardous wastes is regulated; however, biotransformation or biodegradation processes, which are used for resource recovery and reuse purposes, are virtually excluded from federal oversight. If a compound listed in Part 261, Subpart D as a hazardous waste is handled so that it will be beneficially used, re-used, recycled, or reclaimed, then the management of that hazardous compound is only subject to transportation and storage requirements.

Table 2.   Clean-up Standards: A Glossary of Terms

Low Level Nuclear Wastes

Below Regulatory Concern (BRC). A limit, established by NRC, on the concentration of radioisotopes that is allowable. Standards and procedures established by the NRC that exempt wastes containing isotopes from regulation and from disposal as a low level waste when the radionuclides are very low in concentration.

Radionuclides. For nonworker exposure from low level waste disposal facilities, the Nuclear Regulatory Commission establishes a 25 millirems performance standard for such facilities as the maximum annual whole body dose (10 CFR 61). The EPA is in the process of issuing standards for releases of radionuclides into the air and groundwater.

Hazardous Waste Program

RCRA Maximum Concentration Levels. Standards adopted for 14 toxic compounds, primarily toxic metals and pesticides, as part of the RCRA groundwater protection standards (40 CFR 264.94).

Water Program

Water Quality Standards. Determinations made by the states of the uses to be made of particular water bodies and the limits on each pollutant necessary to achieve and protect their uses. These include water quality (human health) and ambient water quality (protection of aquatic life) criteria. (The Gold Book: *Quality Criteria for Water 1986*, EPA 44/5-86-001, also published in the 51 *Federal Register* 43665).

Effluent Limit. A limit established by the EPA on the amount of a specific pollutant that municipal sewage treatment plants and industrial facilities are allowed to discharge in their effluent or wastewater. Also called a discharge limit.

National Pollutant Discharge Elimination System (NPDES). A program mandated by Section 402 of the *Clean Water Act* under which the EPA established limits on the amounts of specific pollutants that may be discharged by municipal sewage treatment plants and industrial facilities. These limits are incorporated in NPDES permits.

Maximum Contaminant Levels (MCLs). Enforceable health-based standards set by the EPA that apply to specific pollutants. Cost and feasibility are taken into account when establishing MCLs. For MCLs, 83 contaminants were identified in the Safe Drinking Water Act amendments of 1986. These included benzene, Aldicarb, PCBs, xylene, toluene, selenium, and PAHs.

Air Program

Lowest Achievable Emission Rate (LAER). A stringent level of pollution control required by the *Clean Air Act* for new or modified industrial facilities in nonattainment areas. The LAER is defined as either the most stringent emission limitation contained in the implementation plan of any state for a category of sources, or as the most stringent emission limitation achieved in practice within an industrial category. In theory, the LAER should be lower than new source performance standards.

National Ambient Air Quality Standards (NAAQS). The *Clean Air Act* requires the EPA to set national ambient air quality standards for six common and widespread outdoor air pollutants: sulfur dioxide, carbon monoxide, particulates, photochemical oxidants, nitrogen dioxide, and lead. "Primary" standards must protect human health with a margin of safety, while "secondary" standards are to protect soil, water, crops, visibility, etc. The act requires the standards to be set without regard to compliance costs.

New Source Performance Standards. Minimum federal emissions limits set by the EPA for all new or substantially modified sources in major polluting industries. The standards are based on the best technology currently available, taking costs into account.

Prevention of Significant Deterioration. In the 1977 amendments to the *Clean Air Act*, congress mandated that areas with air cleaner than required by the National Ambient Air Quality Standards be protected from significant deterioration.

3.7 Type of Organism

Indigenous naturally occurring microorganisms present at the site are not regulated under environmental laws; however, additions of nutrients or electron donors may require approval by a governmental agency. Currently, federal regulations do not cover the environmental applications of nonindigenous, naturally occurring microorganisms for hazardous waste remediation. Although not applicable to bioremediation, the USDA requires commercial products containing naturally occurring microorganisms to be salmonella-free and coliform negative when used in food processing facilities. In addition, the FDA addresses microorganisms that may become indirect or direct food additives. The Plant Protection Act also administered by the USDA can apply to biology-based waste treatment using naturally occurring soil organisms. Microorganisms that have been genetically altered by man are also covered under TSCA, USDA/APHIS rules, and the National Institute of Health (NIH) Field Release Guidelines.

4. EXAMPLES OF REGULATIONS WHICH CREATE MARKETS

4.1 Subtitle I of RCRA

The regulation of leaking underground storage tanks addressed under RCRA has benefitted the bioremediation industry by creating new markets. On September 23, 1988, the final technical performance standards and associated regulations implementing the federal UST program were published. The program, unique to RCRA, applies to products as well as wastes. Subtitle I of RCRA addresses underground tanks storing regulated substances including petroleum products (e.g., gasoline and crude oil) and Superfund-defined hazardous substances. Tanks storing hazardous wastes are regulated under Subtitle C of RCRA. Clean-up standards have not been promulgated for the federal UST program; instead, the remediation level is negotiated for each site. The restoration of soils and groundwater contaminated by underground storage tanks is a major market for the biotreatment industry and has grown substantially in the last few years because of this absence of national remedy standards and the ease of petroleum product biodegradation.

4.2 Clean Water Act

Water quality standards for discharging wastes has benefitted biotechnology. Any portion of a RCRA regulated TSD that discharges wastes into navigable waters must obtain a National Pollutant Discharge Elimination System permit under the CWA. In addition, any action by a TSD that discharges hazardous wastes into a sewer line feeding into a POTW must comply with the CWA's pretreatment standards. It is the pretreatment program established to meet water quality standards that will create a sizable market for the application of biological treatment of dilute hazardous waste. As in the RCRA LDRs, the levels of clean-up (residual concentration of the pollutant in the effluent) specified by the agency and/or specific treatment methodologies will dictate whether bioremediation can be used. Congress has designated which pollutants will be addressed under the CWA program. In addition, congress identified 83 pollutants (radioactive sub-stances, bacteria, viruses, volatile organic chemicals such as benzene, organic chemicals such as PCBs, and inorganic chemicals such as cyanide) in its latest amendments to the Safe Drinking Water Act. BAT is to be identified for each contaminant.

4.3 Clean Air Act

Air quality standards have also benefitted biotechnology. Emissions coming from a RCRA TSD, e.g., incinerator or land disposal facility, must meet the performance standards set out by the CAA. Moreover, hazardous pollutants emitted from medical waste incinerators and municipal sanitary landfills may also be required under federal or local air rules to meet minimum discharge standards.

Biology-based waste treatment offers two approaches to the control of toxic air emissions. The first approach creates a market for environmental biotechnology: the treatment of the waste before burning. Desulfurization of coal and sweetening of sour petroleum are good examples of that application. Treatment of the incinerator's or landfill's emissions would be the second application for bioremediation. The major pollutant emitted from incinerators would be metals or radioactive compounds, since most organics would be destroyed by the combus-tion process. A landfill, on the other hand, may emit methane or other volatile organics. Microorganisms could trap the metals or isotopes from the incinerator, while bacteria, et al. could degrade the volatiles from a landfill into primary elements and cell mass.

The use of biodegradation to decrease air emissions, however, is only one side of the impact the CAA may have on biological treatment. Biotreatment may be the source of emissions. A procedure routinely used to introduce oxygen to the microbial consortia is mixing the contaminated soil or water intentionally expos-ing a large surface to ambient air. If the pollution consists of volatile organics, this tilling may disperse the contaminants into the air about the biological treatment unit. At least one state (Michigan) does not permit open air biological treatment systems for that reason. To minimize volatilization, containment structures may be in order for some environmental biotechnology applications.

5. CONCLUSION

In summary, laws and regulations can create a demand for a particular waste treatment technology. The demand for biology-based waste treatment has increased in the last three years, both based on statutory language encouraging permanent waste treatment methodologies and on U.S. EPA policy bolstered by an agency/university/Department of Energy (DOE)/Department of Defense (DOD)/public/industry bioremediation action committee.[15]

Because the regulatory development process is slow, the positive attitude about biodegradation reflected by the U.S. EPA's managers is only now beginning to appear in the regulations. Environmental regulations, at this juncture, have not been written with a biological approach in mind. For example, when a contaminated site is surveyed, data are not routinely gathered on the indigenous microbial population, on trace nutrients necessary for biomass growth, or on bioavailability of the pollutants, to name a few parameters important to biodegradation. The land disposal restrictions were promulgated before the regulators were aware that environmental biotechnology was a possible treatment regime. The proposed corrective action regulations recognized the need for clean-up levels different from those set in the LDRs when remediating historically contaminated soils and debris. Moreover, the agency has initiated a separate rule-making process for such soils.

However, these regulatory "fixes" are not yet in place. For the short term, rules that have been written for other treatment processes (such as incineration) may slow down the applications of bioremediation in the field. To compound this potential, the hazardous waste regulatory net is expanding. For example, selected petroleum refinery wastes were listed effective May 2, 1991, as RCRA Subtitle C wastes, and others are in the process of being included. Under the CAA, rules are under development that set air emission standards for RCRA treatment, storage, and disposal facilities and for selected volatile chemicals such as hazardous organics, perchloroethylene, and organic solvent degreasers. Under the Toxic Substances Control Act (TSCA), in addition to an upcoming proposal for microbial products of biotechnology, a rule is under development that sets a minimum soil concentration for dioxins and furans. A final rule under the CWA is nearing completion for sewage sludge use and disposal. The CWA effluent guidelines are at the proposed rule stage covering pharmaceutical, pesticide, and pulp, paper, and paperboard categories.

In closing, the use of microorganisms in hazardous waste management is governed by a myriad of complicated laws, regulations, and policies. The purpose of this chapter was to demonstrate the variety and complexity of governmental programs that may control biology-based waste treatment and to alert the reader to their existence. Overlap and gaps among rules exist. For that reason, as in all environmental programs, the ultimate decision of which rules and procedures are appropriate for a particular activity lies with the federal and state regulators. A good working relationship with these individuals is imperative if bioremediation is to become commonplace as a field treatment methodology.

6. REFERENCES

1. Molnaa, B. Formerly of Solmar Corporation, Orange, CA, 1989.
2. "Points to Consider in the Preparation and Submission of TSCA Notifications for Microorganisms," Office of Toxic Substances, U.S. EPA (December 5, 1989).
3. The purpose of the "Coordinated Framework" was to specify particular federal agencies as being the regulatory lead for categories of biotechnology activities and biotechnology-derived products. The document was jointly published by the Office of Science and Technology representing the White House, the Environmental Protection Agency, the Food and Drug Administration, the Department of Agriculture, the Labor Department (Office of the Safety and Health Administration), and the National Institutes for Health.
4. Markland Day, S. "Federal Regulations: How They Impact Research and Commercialization of Biological Treatment," in *Environmental Biotechnology for Waste Treatment*, G. S. Sayler, R. Fox, and J. Blackburn, Eds. (New York: Plenum Press, 1991), pp. 217–232.
5. "Research and Special Programs Administration: Etiologic Agents," (49 CFR Parts 172 and 173), *Federal Register*, Vol. 56, No. 2 (January 3, 1991).
6. The mixture rule defines any mixture of a solid waste and a listed hazardous waste as a hazardous waste.
7. The RCRA addresses solid wastes, while the CWA controls the management and treatment of contaminated water and the CAA, polluted air.
8. As illustrated by the chapters in this book, organics are a prime target for biotechnology mediated mineralization. On March 29, 1990, the "Hazardous Waste Management System: Identification and Listing of Hazardous Waste; Toxicity Characteristics Revisions" was published as a final RCRA rule. This rule replaces the Extraction Procedure Leach Test with the Toxicity Characteristics Leaching Procedure (TCLP), adds 25 organic chemicals to the Subtitle C regulatory universe, and establishes regulatory levels for these organic chemicals.
9. The PCB's are not covered under the RCRA Treatability Exemption because they are regulated under the TSCA.
10. "*Bioremediation in the Field*," Office of Solid Waste and Emergency Response and Office of Research and Development, U.S. EPA (November, 1991) describes the use of a no-migration exemption for a land treatment unit. Under the LDRs, contaminated soils cannot be moved from the pollution source and placed in a land-based unit unless they have been treated to a promulgated clean-up level or by a specified technology.
11. The permit exemption for the federal Superfund program was specified by the statute in the Superfund Amendments and Reauthorization Act (SARA) of 1986.
12. "*CERCLA Compliance With Other Laws Manual: Interim Final*," EPA/540/G-89/006 (August 8, 1990).
13. Day, Sue Markland. "Federal Regulations: Are They Helping or Hindering Bioremediation?" 1992 Spring National AICHE Meeting, New Orleans, LA (1992).
14. PL 101-549, Section 118 (a): Control of Pollution from Federal Facilities.

15. "Summary of Report on the EPA-Industry Meeting on Environmental Applications of Biotechnology," U.S. Office of Research and Development EPA, Crystal City, VA (February 22, 1990).

7. GENERAL REFERENCES

7.1 Laws

- The **Resource Conservation and Recovery Act of 1976 (PL 94-580)** amended the Solid Waste Disposal Act of 1965 (PL 89-272) to create a cradle to grave regulatory system for hazardous waste. The act requires generators, transporters, and disposers to maintain written records of waste transfers and requires the EPA to establish standards, procedures, and permit requirements for disposal.
- The Solid Waste Disposal Act was again amended by the **Hazardous and Solid Waste Amendments of 1984 (PL 98-616).** The HSWA amendments toughened standards for land disposal facilities, banned land disposal of many highly toxic wastes, required regulations of businesses that generate small quantities of hazardous waste, and authorized a regulatory program for underground storage tanks for petroleum and hazardous substances.
- Amending the Comprehensive Environmental Response, Compensation and Liability Act of 1980 (PL 96-510) was the **Superfund Amendments and Reauthorization Act of 1986 (PL 99-499).** This act, as amended, gave the EPA power to force those responsible for hazardous waste sites or other hazardous substance releases to conduct clean-up or other response actions.
- The major law protecting the "chemical, physical and biological integrity of the nation's waters" is the **Federal Water Pollution Control Act of 1972 (PL 92-500), amended by the Clean Water Act of 1977 (PL 95-217), and Water Quality Act of 1987 (PL 100-4).** Under this group of laws, the EPA establishes water-quality criteria to develop water quality standards, technology-based effluent limitation guidelines, pretreatment standards, and a national permit program to regulate the discharge of pollutants.
- The **Safe Drinking Water Amendments of 1986 (PL 99-339),** which amended the Safe Drinking Water Act of 1974 (PL 93-523), required the EPA to issue standards within 3 years for 83 contaminants including radioactive substances, bacteria, viruses, volatile organic chemicals such as benzene, organic chemicals such as PCBs, and inorganic chemicals such as cyanide. The act requires drinking water standards to be based on best available technology taking costs into account.
- The EPA is required to review the health and environmental effects of new chemicals and chemicals already in commerce under the **Toxic Substances Control Act of 1976 (PL 94-469).** If a chemical's manufacture, processing, distribution, use, or disposal would create unreasonable risks, the EPA can regulate or ban it. PCBs, asbestos, and radon are covered by this act, as is the deliberate release of genetically engineered microorganisms.
- First passed in 1963, the **Clean Air Act** was rewritten by the **Clean Air Act of 1970 (PL 91-604).** Congress made major revisions and additions by enacting the **Clean Air Act Amendments of 1977 (PL 95-95)** and the **Clean Air Act Amendments of 1990 (PL 101-549).** The act requires the EPA to set national ambient air quality standards for common and widespread pollutants. To achieve the standards, states and the EPA require industries, businesses, and motor vehicles to reduce emissions.

Separate requirements apply to clean air and dirty air areas. The act also establishes programs to control acid rain and toxic air pollution and to protect the stratospheric ozone layer.

- Signed into law on January 1, 1970, the **National Environmental Policy Act of 1969 (PL 91-588)** requires the federal government to develop environmental impact statements before undertaking any major federal actions.
- **Federal Insecticide, Fungicide, and Rodenticide Act of 1947, as amended in 1988 by PL 100-532,** was enacted to protect farmers from ineffective and dangerous pesticides. Registration of all pesticides before marketing is required. Registration is based on a determination of what uses are safe and what restrictions on use are necessary.

7.2 Regulations

Resource Conservation and Recovery Act

"EPA Technical Standards and Corrective Action Requirements for Owners and Operators of Underground Storage Tanks," (40 CFR 280), *Federal Register*, Vol. 53, (September 23, 1988), p. 37194.

"Hazardous Waste Management System: Identification and Listing of Hazardous Waste," Toxicity Characteristics Revisions, U.S. EPA (40 CFR Parts 261, 264, 265, 268, 271, and 302), *Federal Register*, Vol. 55, No. 61 (March 29, 1990).

"Land Disposal Restrictions for Third Scheduled Wastes," Rule, Part II, U.S. EPA (40 CFR Part 148 et al.), *Federal Register*, Vol. 55, No. 106 (June 1, 1990).

"General Pretreatment and National Pollutant Discharge Elimination System Regulations," Final Rule (40 CFR Parts 122 and 403), *Federal Register*, Vol. 55, No. 142, (July 24, 1990).

"Corrective Action for Solid Waste Management Units at Hazardous Waste Management Facilities," Proposed Rule, Part II, U.S. EPA (40 CFR Parts 264, 265, 270, and 271), *Federal Register*, Vol. 55, No. 145 (July 27, 1990).

"Petitions to Allow Land Disposal of a Waste Prohibited under Subpart C of Part 268," (40 CFR Part 268.6).

"Land Disposal Restrictions: Potential Treatment Standards for Newly Identified and Listed Wastes and Contaminated Debris," Advanced Notice of Proposed Rule-making, Office of Solid Waste, U.S. EPA, *Federal Register*, Vol. 56, No. 104 (May 30, 1991).

"Hazardous Waste Treatment, Storage, and Disposal Facilities; Organic Air Emission Standards for Tanks, Surface Impoundments, and Containers," Proposed Rule, Office of Solid Waste, U.S. EPA, *Federal Register*, Vol. 56, No. 140 (July 22, 1991).

Comprehensive Environmental Restoration, Compensation, and Liability Act

"National Oil and Hazardous Substance Pollution Contingency Plan," U.S. EPA (40 CFR Part 300), *Federal Register* (March 8, 1990).

Safe Drinking Water Act

"National Primary Drinking Water Regulations — Synthetic Organic Chemicals and Inorganic Chemicals; Monitoring for Unregulated Contaminants; National Priority Drinking Water Regulations Implementation; National Secondary Drinking Water Regulations," Final Rule, U.S. EPA, *Federal Register*, Vol. 56, No. 20 (January 30, 1991).

Department of Transportation

"Research and Special Programs Administration: Etiologic Agents," (49 CFR Parts 172 and 173), *Federal Register*, Vol. 56, No. 2 (January 3, 1991).

Toxic Substances Control Act

"Disposal of Polychlorinated Biphenyls," Advanced Notice of Proposed Rule-making, Office of Toxic Substances, U.S. EPA, *Federal Register*, Vol. 56, No. 111 (June 10, 1991).

National Institute of Health

"Notice of Actions Under the NIH Guidelines for Research Involving Recombinant DNA Molecules," NIH, Department of Health and Human Services, *Federal Register*, Vol. 56, No. 138 (July 18, 1991).

National Environmental Policy Act

"The Environmental Protection Agency's National Environmental Policy Act Review Procedures for Public and Other Federal Agency Involvement and for the Office of Research and Development," U.S. EPA, *Federal Register*, Vol. 56, No. 87 (May 6, 1991).

7.3 Government Guidances and Policy Papers

MEMORANDUM date October 3,1985 SUBJECT: Research, Development, and Demonstration (RD & D) permits; FROM: Marcia E. Williams, Director of the Office of Solid Waste draft guidance.

"Permit Guidance Manual on Hazardous Waste Land Treatment Demonstrations," Final Version, Office of Solid Waste, U.S. EPA (July, 1986).

"Domestic Sewage Study," EPA/530-SW-86-004.

"The Layman's Guide to the Toxic Substances Control Act," EPA/560/1-87-011.

U.S. EPA, Office of Solid Waste. *RCRA Orientation Manual*, EPA/530-SW-86-001, p. III-23.

"Overview of the Land Disposal Restrictions," Office of Solid Waste, U.S. EPA (September 28, 1989).

"Points to Consider in the Preparation and Submission of TSCA Notificatioı organisms: Environment Release," Issued by the Office of Pesticides and Toxiς (December 5, 1989).

"The Nation's Hazardous Waste Management Program at a Crossroads," Officε of Solid Waste and Emergency Response, U.S. EPA/530-SW-90-069 (1990).

"Biotechnology: Request for Comment on Regulatory Approval," (54 FR 7027), issued by OPTS (February 15, 1990).

"Principles for Federal Oversight of Biotechnology: Planned Introduction into the Environment of Organisms with Modified Hereditary Traits," Office of Science and Technology Policy, *Federal Register*, Vol. 55, No. 147 (July 31, 1990).

7.4 Congressional Publications

"The Clean Water Act as Amended by the Water Quality Act of 1987," Senate Committee on Environmental and Public Works, S. Prt. 100-91, Public Law 100-4, U.S. Government Printing Office (March 1988).

7.5 Federal Regulations That Apply to Genetically Engineered Microorganisms

- **TSCA Rules** (40 CFR Parts 700, 720, 721, 723, and 725, draft rule being circulated: Microbial Products of Biotechnology: Proposed Regulation under the Toxic Substances Control Act) — The production and commercial applications of "new" microorganisms require notification and approval by the Office of Toxic Substances.
- **NIH Guidelines** (FR Vol 51., No. 88, May 7, 1986) — The deliberate release of any organism containing recombinant DNA into the environment requires specific approval by the National Institutes of Health. Experiments in this category cannot be initiated without submission of relevant information on the proposed experiment to NIH, the publication of the proposal in the *Federal Register* for 30 days of comment, review by the Recombinant DNA Advisory Committee (RAC), and specific approval by NIH. These guidelines are applicable to all recombinant DNA research financed by NIH and other participating federal agencies.
- **APHIS** (7 CFR Parts 330 and 340, final rule issued June 16, 1987: Introduction of Organisms or Products Altered or Produced through Genetic Engineering Which are Plant Pests or Which There is Reason to Believe are Plant Pests) — This rule requires a permit for the movement (and/or introduction) of both genetically engineered and nongenetically engineered organisms, products, or selected articles that are plant pests or could harbor plant pests.

Index

223